# A BACKGROUND TO
# ENGINEERING DESIGN

PETER POLAK
*Department of Mechanical Engineering,
University of Sheffield*

*First published 1976 by*
THE MACMILLAN PRESS LTD
*London and Basingstoke*
*Associated companies in New York Dublin*
*Melbourne Johannesburg and Madras*

SBN 333 18771 7

*Text set in 10/11 pt. IBM Press Roman,*
*printed by Photolithography*
*and bound in Great Britain at*
*The Pitman Press, Bath*

# Contents

## 8. Bearings and Seals 65

Deals mainly with rolling and sliding bearings (hydrodynamic and hydro-static types) and fluid seals.

## 9. Damping, Mountings and Vibration 79

Discusses mountings, vibrations, damping, etc., and briefly refers to noise.

## 10. Some Points on Manufacture and Appearance 86

Considers design aspects of manufacturing methods, briefly describing some of the less familiar ones and concludes with a short section on visual design aspects.

*Appendix* 100

List of symbols; Twist—bend buckling; Unsymmetrical sections; Behaviour of bolted joints; Bellows expansion joints; Some useful theorems.

*References* 111

*Index* 115

# Preface

'Design' is a popular expression with varying implications: wallpaper design
differs from dress design, 'industrial design' differs from engineering design. Some
journals go so far as to use the word design for the external aspects of a machine,
calling the insides, a little airily, the 'engineering'; for example, a teleprinter was
described as designed by a designer particularly well known for elegant table-
ware, the makers of the works were mentioned a long way further down in the
'credits'. Engineers in turn insist that the word should refer almost entirely to the
works, or in simple cases to the stressing.

This semantic difference is best resolved by those engineers who themselves
display a strong sense for the appearance of a product and are prepared to recognise that those trained mainly in the visual arts and relatively free from mechanical
habits can have something refreshing to offer.

The present book is concerned almost entirely with functional aspects; indeed
the author feels that not only in machinery should the externals generally take
their place with all the other considerations; for instance in light-fittings, efficiency
and styling can be at loggerheads. Only where the function is simple and sufficient
strength easily provided, as in furniture, can external design take precedence.
Occasionally the function and ease of making actually suggest a happy shape.

Engineering design at its most restricted is finding the right thickness for a part
when shape, function, loading and material are pre-decided. As will be seen later,
even this is not always easy. Higher levels of elaboration are reached in 'shopping-
list' design, finding the most economical and/or versatile process plant or production line consisting of standard but expensive items of equipment. The most
creative design activity, starting from basics, is also the most demanding if it is
not to consist of repeating old mistakes along with inventing a few new ones.

This book does not supersede any established manuals but brings together key
points from the past and more recent data. The references include items dating
back seventy years but the majority are only a few years old.

To avoid items like 'this is a bolt, this is a clamp, this is a keyway', familiarity
with the names and uses of basic mechanical components is assumed, as is a knowledge of basic stress and strain relations appropriate to first- or second-year under-
graduates in engineering.

To reduce tediousness, extensive explanations have been avoided; it is felt that
the intended reader is better served by erring on the side of brevity, leaving room
for thought yet, it is hoped, no room for misunderstanding.

Standard drawing conventions have been varied slightly in the interests of clarity; each figure should be considered independently. Generous use of shading, though unnecessary for many, should help to minimise uncertainties for those students who are unfamiliar with machinery details.

To avoid irritating brackets, SI units are used generally, though occasionally Imperial units are shown. For stresses, a convenient set of figures to relate the more common units together is the ultimate strength of a low-carbon steel, 28 tons/in.$^2 \approx 400$ MN/m$^2 \approx 60\ 000$ p.s.i. ($=$ lbf/in.$^2$).

Apologies for digressions, approximations, simplification of facts and verbal short-cuts are tendered here and now, in bulk.

The author gratefully acknowledges his debt to all those who taught him or provided opportunities for learning and experience.

Thanks are due to Sheffield University for the help received from the Applied Science Library, the photographic section, the workshops and laboratories of the Mechanical Engineering Department and for a certain amount of clerical assistance; also to Sheffield City Library which was found to form a most useful complement to the University library.

# 1
# Introduction and some Ergonomics

Design is not quite the same activity as inventing. New design elements can come from scientific discoveries, from patient experimentation, from chance observations relevant to a current need. Examples of each are the transistor, the electric light bulb, the Bessemer converter and the Mannesmann tube-piercing method. In the case of the electric light bulb the design concept came first, derived from logical use of scientific principles, followed by tests on likely filament materials to find which had the most suitable combination of properties.

Bessemer's converter came about by the opposite sequence. When melting down some pig-iron Bessemer noticed an unmelted pig at the edge of the hearth; prodding it with an iron rod he found it was hollow. From his knowledge of the melting points of pig iron and steel he suspected that the outer skin of this piece had been converted to steel by the flames preferentially removing the carbon, as in the older Osmond-iron process. Further investigations enabled him to devise an apparatus to carry this out in an organised manner.

Mannesmann's method is said to have started over dinner; while waiting for his soup to cool and rolling a piece of bread between his fingers he unexpectedly produced a hole, the high shear gradient at the centre of the oval producing a fracture. He had the imagination to suppose that the same might apply to steel and to devise (invent and design) the crossed-roll system which imitates the finger action on a continuous basis.

We cannot all be great inventors, nor do we need to be; the last two hundred years have provided a vast reservoir of ideas freely available for the betterment of mankind, leaving us a task less glorious but no less important and no less difficult of selecting elements for our purpose. This purpose, need it be said, is to serve the public; it usually should also include profit for our employer. Good design achieves both.

The basic parts of a design can be summed up as strength, function and economy. These, taken in a broad sense, embrace all the features discussed in

1

this book and many more besides which could not be adequately discussed in one small volume intended for students. Since it is hoped that the reader will refer to this book again in later life, a more detailed list of aspects is set out in figure 1 by way of acknowledging their importance and providing a frame for those discussed in some detail.

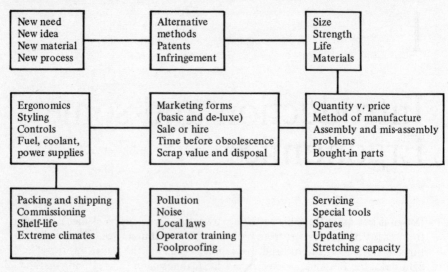

| New need<br>New idea<br>New material<br>New process | Alternative<br>methods<br>Patents<br>Infringement | Size<br>Strength<br>Life<br>Materials |
|---|---|---|
| Ergonomics<br>Styling<br>Controls<br>Fuel, coolant,<br>power supplies | Marketing forms<br>(basic and de-luxe)<br>Sale or hire<br>Time before obsolescence<br>Scrap value and disposal | Quantity v. price<br>Method of manufacture<br>Assembly and mis-assembly<br>problems<br>Bought-in parts |
| Packing and shipping<br>Commissioning<br>Shelf-life<br>Extreme climates | Pollution<br>Noise<br>Local laws<br>Operator training<br>Foolproofing | Servicing<br>Special tools<br>Spares<br>Updating<br>Stretching capacity |

Figure 1

A design generally relies on some proportion of bought-in equipment which needs to be understood sufficiently to ensure suitable selection and satisfactory performance. This aspect is well illustrated by discussing electrical equipment. The design work which has brought such equipment to its high state of efficiency is beyond the scope of a general book such as this, yet the general designer should be aware of the kinds of aspect which follow.

We tend to forget about the high starting torque of electric motors, especially where direct on-line starting can be employed to avoid expensive starting-gear. Usually this is safely absorbed in accelerating the motor itself, but if the load has high rotary inertia much of the extra torque may get through and allowance must be made for it. Sometimes it pays to use fluid- or powder-filled couplings in the drive-line. These permit considerable slip during run-up but give little or no loss at full speed.

Frequent starting gives rise to extra heating in the motor windings and also in slip-type couplings; this must be considered in selecting the right size.

In mobile equipment, motor weight may be important. The power output of electric motors is related to the number of flux changes; hence high-frequency motors and high-speed commutator motors tend to have much higher power—weight ratios than 50—60 Hz induction motors but require high-frequency generators in the former case and may be noisy in the latter.

In complex plant, sequential starting is provided for so that essential services such as controls, lubricants and coolants come on before the main plant. Starting

2

gear is usually designed to trip out if power is interrupted, but consider the effect of a momentary stoppage: if the main starter is large, its inertia may let it stay 'on' while the ancillaries trip out. This can be prevented by suitable wiring.

Troubles can arise from poor contact design, especially with frequent operation or at low voltage. A contact mechanism needs enough resilience to maintain a good contact force despite wear, thermal expansion or creep; it should include some sliding action to displace oxide films and should open and close rapidly to minimise arcing. The contact breaker points on a car meet these requirements; they are spring-loaded to close and have a small amount of sliding motion — a large amount would cause excessive wear in view of the very large number of operations.

Small switches employ over-centre spring action or a permanent magnet to prevent dither during opening or closing. Sliding motion is obtained by flexure of the contact carriers and their angular position. In the micro-switch shown in figure 2a, pushing the button moves the V-notch pivot below the line of the tension spring whereupon the blade snaps across to the dashed position. In another form of micro-switch the whole mechanism is pressed from one piece of metal, and ingeniously placed corrugations put some parts under initial tension, the rest in compression.

The thermostat switch shown in figure 2b uses the principle that magnets give increasing attraction as the air-gap is reduced and greater flux is produced, and vice versa, to give snappy closing and opening. The same effect also provides a lag or dwell to avoid over-frequent operation at small temperature changes. The flexure of the lower contact is shown grossly exaggerated to demonstrate the sliding motion.

The manual knife-switch in figure 2c shows an auxiliary contact spring-loaded to snap out rapidly, breaking circuit after the main blade. On closing, again the auxiliary contact takes the brunt of any arcing.

These points are not only intended to be directly useful; they also indicate the type of attitude which should be applied to all bought-in equipment.

Turning now to ergonomics we require basic data, such as the material collected by Woodson and Conover[1], together with either an exceptional degree of imagination or a liberal use of mock-ups. Please remember also that not all operators are of the same size and strength as the median U.S. male. Some are small and nimble yet may be quite strong, others are large and very strong indeed.

In considering strength, we distinguish between occasional use of maximum force, repetitive work cycles and intermediate cases such as loading batches of goods on to trucks. It is salutary to remember that an output of 120 W propels an average adult up a one-in-ten slope at marching pace. Athletes know that for most muscular actions there are optimum speeds for endurance; it seems to help if the speed can be varied occasionally. The reader may like to view muscles as similar to hydraulic cylinders with leakage and friction. For a given work-rate, leakage causes the greatest loss at low speeds while friction predominates at high speeds. Optimum efficiency is somewhere in between.

We tend to forget Newton's law of action and reaction. Pushing a car on snow or carrying a load uphill in backless beach-shoes are obvious examples, but try sawing while sitting on a stool. Many recommendations on working positions

3

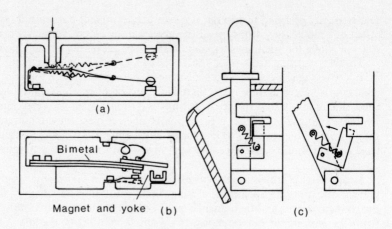

(a)

Bimetal

Magnet and yoke  (b)

(c)

Figure 2. Switches (schematic)

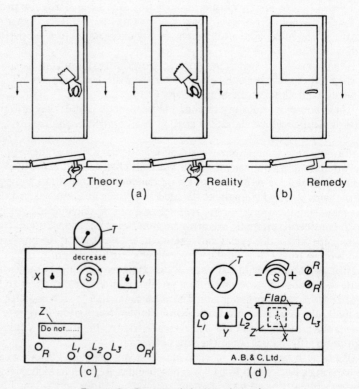

Theory  (a)  Reality  Remedy  (b)

( c )  ( d )

Figure 3. Poor and better layouts

4

seem to come from those wielding nothing heavier than a pencil; they should try working on their layouts at full force for a thirty-five hour week. Even when relaxed in a chair, leaning back can produce a forward reaction dragging one's skirt up: seat designers please note.

Musicians and singers like to sit and stand alternately and walk about during intervals, realising that there is no single comfortable working position; a good work-plan should encourage variety of position and movement. Walking about to collect more material, bending down occasionally to sweep up waste, etc., can reduce fatigue; if the designer makes this possible without loss of machine time by building in safety-trips, etc., so much the better.

Hand controls should be easy to learn, following established conventions, easy to distinguish by size and detailed shape, and resistant enough not to be moved by catching on clothing. Start buttons should be recessed, stop buttons protruding, big and red. In Europe there is an unfortunate inconsistency in electrical work: switches are usually up when off whereas isolators for machines are down when off, perhaps to make the safety interlock slightly easier to design.

A particularly treacherous control layout was reported some years ago from the aircraft industry. One aircraft had the propeller pitch controls in a position often used in other types for the engine throttles; the movement for reducing pitch was in the direction usually associated with throttle opening. A pilot trained on other machines might, while intending to put on more power, inadvertently reduce pitch; this would allow the engines to speed up audibly as expected but instead of getting the increased thrust and lift required, the opposite would happen. If the extra power was needed urgently, this could be disastrous.

Foot controls should be in line with the operator, otherwise lateral reaction can cause the foot to slip off. Pedals are often unsatisfactory in shape, size and finish and merit closer attention; slightly concave designs tend to discourage slipping off. If foot controls are to be used standing, a hand-hold should be provided.

Designers of small hand-tools seem unaware of the shape of the closed hand. This varies with the size of the object to be grasped; the natural shapes are the flat bar, triangle and square, at the very most the pentagon. These also give some hint of the angular position. The usual round or pseudo-round handle simply encourages blisters. On the other hand, heavier tools such as picks and axes are usually excellently designed.

Now follows a description of some actual poor layouts and suggestions for improving them. Figure 3a shows a door-fitting recently installed in a busy public building. The clearance between door knob and fixed frame is just about tolerable when the knob is upright; when turned, the gap is just less than finger thickness. The remedy here is obvious — fit a lever instead of the knob as shown in Figure 3b.

Figure 3c shows a control panel. The first switch from the left, $X$, must not be switched on until the incoming supply and the main switch $Y$ have been on for five minutes. The instruction to this effect is tucked away in a corner at $Z$. A row of lights with somewhat ambiguous labels is placed at $L_1$, $L_2$ and $L_3$. Below the tachometer $T$ is a speed-control knob $S$ which is turned anti-clockwise to increase speed, the opposite way to usual quantity controls and also to the tachometer movement. $R$ and $R'$ are range-adjusters which affect the stability of the system. They protrude and turn so easily that by leaning against the console one can displace the setting. Please note that this is not an imaginary example; the author uses this panel frequently.

A better scheme is shown in figure 3d. The tachometer $T$ is to the left of the speed-control $S$ which is better for right-handed operators; the speed-control is turned clockwise to increase speed. The range adjusters $R$ and $R'$ are recessed buttons with screwdriver slots. Below, $L_1$ indicates incoming power, switch $Y$ comes next, turning on the supply to the machine and also to light $L_2$. Next comes the critical switch $X$, covered by a hinged flap with the instructions $Z$ placed on the flap so they cannot be overlooked. This is followed by light $L_3$ which shows when everything is operational.

# 2

# Loads and Structures

One of our first duties in design is to make things strong enough. Accepting this, the most efficient structure is that in which the material is strategically placed so as to give both the lightest structure for a given load and permissible stress level and also the stiffest structure for a given weight.

Such a structure can be quite complex and expensive to make; the closest approach to the ideal is justified in aero-space and in long-span bridges. In the former, lightness brings enormous benefits since the payload is small compared with structure and fuel. In the latter also the dead-weight can be the major loading so that any weight-saving is highly cumulative. The total equation also includes running costs (such as fuel, labour and interest on capital) and stiffness; good design pays handsomely in every case.

In the design of smaller structures, economy of manufacture causes us to depart further from ideal forms but understanding the ideal in the first place helps us to design well, whether we use ready-made structural elements or design the shapes to suit the job. Structural elements are usually made in uniform sections such as sawn timber, metal joists, tubes, wire ropes, corrugated sheeting, etc. So let us start with a simple post-and-beam structure using a uniform beam, equally strong either way up and carrying a uniformly distributed load. (Most beams are equally strong either way up though not sideways; exceptions are reinforced concrete beams with the reinforcement on the tension side only, crossing over if necessary, and also slim T-shaped beams where the thin leg or rib is strong in tension but may buckle laterally if put into serious compression.) Where is the best place for the supports if free choice is permitted?

Figure 4 shows the same loaded beam with supports first at the ends and second some distance inwards, with the shear-force and bending-moment diagrams for each, to the same scale. The large reduction in bending moment due to bringing the props inwards is obvious. If they are too close together, the overhanging canti-lever gives the largest moment. If they are too far apart, the sagging moment in the middle is the worst. The optimum is when these peak moments are numerically equal. If the support is spread over an appreciable length (shown dotted), or if the beam is not equally strong against sagging and hogging moments, the optimum is a little more difficult to find but still occurs with the supports some distance inwards from the ends.

Referring to figure 4, beam length = $2a$, overhang = $x$. If loading = $w$ per unit length we have, from total equilibrium and symmetry,

$$\text{reaction at each support} = aw$$

$$\text{bending moment at a support} = \tfrac{1}{2}wx^2$$

$$\text{bending moment at centre} = \tfrac{1}{2}wa^2 - aw\,(a - x)$$

(due to half the load and one reaction). To achieve our object

$$|\tfrac{1}{2}wx^2| = |\tfrac{1}{2}wa^2 - aw(a - x)|$$

By inspection, having maintained the same sign convention (clockwise moments positive) and dividing through by $\tfrac{1}{2}w$

$$x^2 = -(a^2 - 2a^2 + 2ax)$$

$$x^2 + 2ax - a^2 = 0$$

$$x = -a \pm \sqrt{(a^2 + a^2)} = 0.414a$$

The existence of peak bending moments at these points was deduced by inspection, though that in the centre could have been obtained mathematically in the usual way. The peaks at the supports would not show up mathematically, being at a discontinuity; however, it is still true that the bending moment comes to a peak where the shear force goes through zero, as usual. Thus the supports should be about one-fifth of the length from each end. The other equation, $\tfrac{1}{2}wx^2$ $= + [\tfrac{1}{2}wa^2 - wa(a - x)]$, gives the answer $x = a$, implying one central support; obviously then the moments are identical, but not economic.

An equally easy case occurs if the beam has negligible weight but carries a moving point load; this gives an overhang of one-sixth. A more complex case is shown in figure 5, a uniform beam and load, hinged at one end, with a choice of propping point. (This is not the propped cantilever beloved by textbook writers; that is an ideal rarely relevant in practice since neither the built-in end nor the prop are rigid and their elasticities are unlikely to cancel each other.) Now we cannot find the point of maximum bending moment by symmetry. It can be done by differentiating the bending-moment equation, but more easily by referring to the shear-force diagram.

Taking moments about the hinge

$$R_2\,(a - x) = \tfrac{1}{2}wa^2$$

$$R_2 = \frac{wa^2}{2(a - x)}$$

$$R_1 = wa - R_2$$

$$= \frac{wa(a - 2x)}{2(a - x)}$$

8

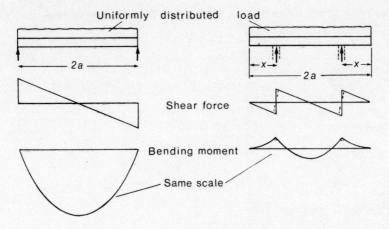

Figure 4. Optimising support positions

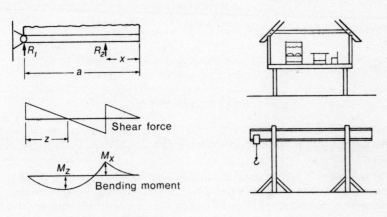

Figure 5. Hinged beam

Figure 6.

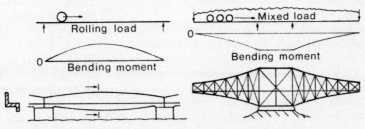

Figure 7. Structure shaped to suit loading

9

The point of maximum bending moment is $z$ from the hinge where

$$R_1 - wz = 0 \qquad\qquad (1)$$

$$M_z = R_1 z - \tfrac{1}{2}wz^2$$

From equation 1

$$R_1 = wz$$

Hence

$$M_z = wz^2 - \tfrac{1}{2}wz^2 = \tfrac{1}{2}wz^2$$

For greatest economy, $M_z = M_x$ and obviously $M_x = \tfrac{1}{2}wx^2$. Therefore

$$x^2 = z^2$$

$$x = \pm z$$

But from equation 1

$$z = R_1/w$$

$$= \frac{a(2-x)}{2(a-x)}$$

Therefore the optimum value for $x$ is

$$x = a\left(1 - \frac{1}{\sqrt{2}}\right) = 0.293a$$

The general idea of using the strength of a beam economically is seen in medieval timber houses with protruding upper floors, and also in some loading-yard cranes (figure 6).

Often we are able to optimise the shape of the structure itself. A simple example occurred on early railways: cast-iron rails spanning between stone blocks were of bowed shape (figure 7). (The bowed 'plate' rail seems to have been invented, either by John Curr or Abraham Darby, before the better-known fish-bellied rail of Jessop.) If the structure is large enough, lattice girders are used, either the Warren girder which is of constant depth but with each member of cross-section appropriate to its load, or structures using uniform flange members with the over-all depth varying to follow the shape of the bending-moment diagram. The bow-string girder is typical of short bridges, while the Forth rail-bridge is a notable example of double cantilevers standing on piers. Suspension and arch bridges are not quite the same as girders; they use the earth to form the lower member, transmitting the forces through massive anchors at the ends.

Mechanical engineers are usually concerned with small spans where often there is choice over the vertical height. Figure 8 shows two simple frames identical in all but height. Below each frame is shown a triangle of forces around a support, from which the forces in the members a and b are found. The force in member c is found by inspection. We observe that the taller structure has longer members but with much less force in them so they can be made lighter. To find the lightest possible structure of the given form, we express the weight of the members in terms of $c$, the length of member c, which is at our choice. The weight of each member is density x length x cross-sectional area. The cross-sectional area is determined either by resistance to buckling, which is dealt with in chapter 3, or simply by the permitted stress in the material, in which case cross-sectional area = force/stress.

Using $\rho$ for density and $\sigma$ for stress, we obtain the following.

| Member | Length | Force | Weight |
|---|---|---|---|
| a | $\sqrt{(b^2 + c^2)}$ | $\dfrac{P}{2} \times \dfrac{\sqrt{(b^2 + c^2)}}{c}$ | $\dfrac{P\rho}{2\sigma} \times \dfrac{(b^2 + c^2)}{c}$ |
| a' | $\sqrt{(b^2 + c^2)}$ | $\dfrac{P}{2} \times \dfrac{\sqrt{(b^2 + c^2)}}{c}$ | $\dfrac{P\rho}{2\sigma} \times \dfrac{(b^2 + c^2)}{c}$ |
| b | $b$ | $\dfrac{P}{2} \times \dfrac{b}{c}$ | $\dfrac{P\rho}{2\sigma} \times \dfrac{b^2}{c}$ |
| b' | $b$ | $\dfrac{P}{2} \times \dfrac{b}{c}$ | $\dfrac{P\rho}{2\sigma} \times \dfrac{b^2}{c}$ |
| c | $c$ | $P$ | $\dfrac{P\rho}{2\sigma} \times 2c$ |

$$\text{Total weight} = \frac{P\rho}{2\sigma}\left(\frac{2(b^2 + c^2)}{c} + \frac{2b^2}{c} + 2c\right)$$

$$= \frac{2P\rho}{\sigma}\left(\frac{b^2}{c} + c\right)$$

To find the minimum weight we differentiate with respect to $c$ and set to zero

$$\frac{d}{dc}\left(\frac{b^2}{c} + c\right) = -\frac{b^2}{c^2} + 1 = 0$$

therefore $$c = b$$

A general guide to this situation is by Michell[2]. His treatment defines a mathematical field which must be filled to obtain the lightest structure. No advantage is gained by going larger; in practice the weight penalty of junctions and the risk of buckling of long thin compression members lead to designs some way inside the Michell field. The theory is valid even if compressive stresses are fixed at some value other than the tensile stresses. The ideal structures for a point load midway between two supports are shown in figure 9 which is taken from Michell. The first structure is best when all space above and below is available (infinite Michell field); the second is preferable when only half is allowed (semi-infinite field). Two simple framed structures are shown for comparison, each separately optimised for vertical height. Note that both these tend to fill up the relevant Michell fields. Further development of this topic may be found in Chan[3] and Owen[4].

11

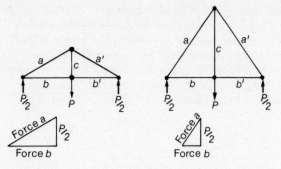

Figure 8. A simple optimising exercise (see text).

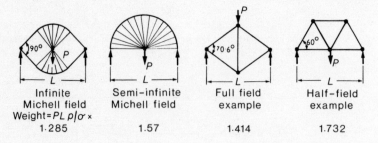

| Infinite Michell field | Semi-infinite Michell field | Full field example | Half-field example |

Weight=$PL\,\rho/\sigma$ ×

| 1·285 | 1.57 | 1.414 | 1.732 |

Figure 9. Michell's and other optimised structures

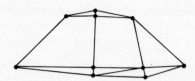

Figure 10. Frame resembling car bodywork

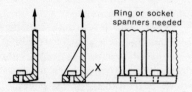

Figure 11. Using webs to stiffen bolted joints

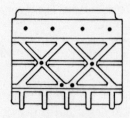

Figure 12. Rigidised engine

It is generally accepted that fully triangulated structures are lightest and stiffest for a given load because they consist of tensile and compressive members only. Members in bending make less efficient use of the material, only the surface layers being fully stressed. When loads in several planes occur the structure should be triangulated in three dimensions, giving what is called a space-frame.

Practical structures often depart from this with more or less justification. Motor-cars correspond somewhat to a framed equivalent shown in figure 10. The central panel has no shear member but the loading is such that there is relatively little shear force here, the floor and roof girders being sufficient. These act as tension and compression flanges for the whole body and also as local beams or shear panels. In coach bodies the window pillars, often very thin, form a very flexible and inefficient connection so that the roof contributes relatively little, however, the deep sides form a sufficient girder and near the end, where doors are placed, forces and moments are low.

In machinery we are often concerned with stiffness; the stresses can be very modest, $1-2$ tons/in.$^2$ or $15-30$ MN/m$^2$ being quite usual. Bolted connections have two problems: deflection of flat flanges, which is overcome by being very generous with gussets, and elasticity of the bolts. This latter is less easily pictured. A small load applied upwards to the joint in figure 11 causes not so much a stretch in the bolt as a readjustment of contact forces. The load is resisted mainly by a reduction of the contact pressure at X which involves relatively little deflection. At overloads, the contact becomes zero and the bolt extends; this gives much larger deflections (less stiffness). If a resilient packing is placed in the joint, the stiffness is greatly reduced even before separation. (This topic is discussed in more detail on p. 103.)

Machinery bed-plates and columns are often subjected to side forces, resulting in twist. Closed tubular or box sections are stiff in twist but channel sections are not; tubes with a longitudinal slit are strikingly less stiff than uncut tubes. This is worth demonstrating: take two cardboard tubes, slit one down its length and compare their torsional stiffness, just by hand. Machine parts cannot always be tubular and bracing may be required to stiffen them. Good lathe beds sometimes contain diagonal bracing, employing the principle of triangulation like a Warren girder[5].

To conclude this chapter, let us consider the structure of a multi-cylinder engine. It has become apparent that of the noise emitted by engines a fair proportion comes from the casing deflecting under the loads. In the past attention was concentrated on the vertical (firing) loads while the side loads were ignored. These give lateral and torsional inputs to the block. We may expect future engine blocks to have flanges and cross-bracing, perhaps as in figure 12, so as to transmit local forces to the main mass with minimal displacements.

# 3

# How Strength can Fail

A major source of confusion for students is the expression 'safety factor'. It may be of some use in a qualitative sense, indicating which factors in a problem tend towards safety and which towards danger of failure. The trouble starts when attempting to give it a numerical value. In design practice there is no single value for assessing safety margins, as can be seen from the examples below.

Gears designed according to BS 436 for any particular length of life under a given normal loading are automatically strong enough to stand occasional overloads of about twice normal load.

In aircraft structures it is recognised that in addition to normal loads severe gust loads will occur, though rarely, and after a certain number of high loads cracks may appear and some parts may need replacing.

In ball-bearings the ratio of load to life is sufficiently well known to allow us to design for statistical probability of survival; the safety margin is then best expressed in running hours rather than load.

The foregoing are examples of limited-life design; in steel buildings and pressure vessels the levels of stress are prescribed by codes to be below yield-point divided by 1.6 to 2 in the straightforward parts, but in the case of some joints the calculated stress must be below about U.T.S./15. (Ultimate tensile strength is given by failure load/original cross-sectional area.)

This should make it clear that since 'safety factors' for the examples above, all taken from good practice, can range from less than 1 up to 15, such a crude factor is sheer nonsense.

Since orthodox stress calculations are very adequately dealt with in existing books, the present volume concentrates on buckling and fatigue problems. In chapter 6 attention is also drawn to brittle fracture at low temperatures, with reference to steel.

The simplest buckling case is the slender column, or Euler strut, discussed in many textbooks. The criticial load for a slender strut is that at which it changes from pure compression to lateral bowing, at constant or even at decreasing force. Figure 13 shows a range of cases together with the formula for various end conditions and load constraints. The critical or crippling load $P_{cr}$ depends on the length

$L$, Young's modulus $E$ and the *lowest* second moment of area of the member (moment of inertia, or $I$-value for short) in bending.

The second case is perhaps unusual in practice but all the rest resemble practical conditions. It is generally considered advisable to keep well away from crippling loads, say by a factor of 4; — even more if the loading is appreciably non-central.

Helical compression springs are also struts and can pop out sideways. Usually they are mounted as line- and angle-fixed, sometimes just angle-fixed. The $I$-value of a helical spring may be found in various textbooks; the short design suggestion is that if a spring is over three diameters long it is best to confine it by a guide, internally or externally.

Although buckling is an elastic phenomenon, in practice there are many cases where plain compressive yielding and buckling failure values become similar, and we may expect some interaction, especially if the compressive stress gets above two-thirds of yield or proof stress; this will be adequately covered if the factor of 4 is used as suggested above.

The second moments of sections are listed in many books, but there is one which is not always easily available. An angle section and its lowest second moment are shown in figure 14. Note that the wall thickness $t$ is restricted: if the wall is thicker the expression becomes inaccurate; if it is much thinner, another danger appears. Consider a flexible steel measuring-tape kept rigid by its transverse curvature, or a metal slat for a Venetian blind. This is much weaker in bending towards the concave side than the opposite side, owing to a form of buckling. The centre of the section is stabilised by its curvature, but the edges less so. Thus if we load the member with the free edges in tension, its strength is much greater than when they are in compression and liable to buckle.

Thin-walled tubes can fail by local buckling of this kind[6]. Here $t << D$, where $D$ is the tube diameter and $t$ is the wall thickness, and in theory failure will occur when the axial stress in compression reaches $2Et/[\sqrt{3(1-v^2)}D]$ (approximately $1.2Et/D$), even when the walls are prefectly straight axially. Waviness brings this down further; a local dimple $\frac{1}{2}t$ deep can bring the value down to $0.54Et/D$.[7]

Cylinders subject to external (radial) pressure are capable of buckling (caving in) when a critical unit pressure $p$ on the outer surface is exceeded. This is analogous to struts, as follows. In a strut a deflection leaves the load line offset, giving a resultant force which increases the deflection further. In a cylinder, the compression in the wall balances the external pressure arch-wise; if the curvature of the arch is not uniform, again an out-of-line effect appears (figure 15). The effect becomes self-sustaining if the pressure is high enough (critical). A long tube will buckle at an external pressure of about $2.2Et^3/D^3$. This is in effect an Euler strut equivalent. If the tube is of length $L$ where $L$ is $5D$ or less and its ends are held circular, it is rather stronger. Extensive details are given in an E.S.D.U. data sheet[8]; typical basic cases can be reduced to the relation $p_{cr} \approx 2.4Et^{2.5}/(LD^{1.5})$ which agrees with expressions going back at least to Southwell in 1915.

Spherical vessels can also buckle, even part-spherical domes. The reader is particularly warned against an expression found in a famous textbook which is over-optimistic by a factor of 6. Von Karman and Tsien[9] give a much better theory which agrees well with experimental values. The critical external pressure is about $0.35Et^2/R^2$ where $E$ is Young's modulus, $t$ the wall thickness and $R$ the radius of the sphere. In practice pressures must be kept well below this. For

15

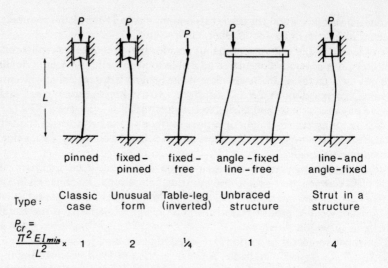

| | pinned | fixed – pinned | fixed – free | angle – fixed line – free | line – and angle – fixed |
|---|---|---|---|---|---|
| Type: | Classic case | Unusual form | Table-leg (inverted) | Unbraced structure | Strut in a structure |

$P_{cr} = \dfrac{\pi^2 EI_{min}}{L^2} \times$  1    2    ¼    1    4

Figure 13. Some critical loads in buckling – keep well away

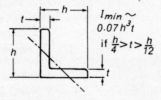

$I_{min} \sim 0.07\,h^3 t$

if $\dfrac{h}{4} > t > \dfrac{h}{12}$

Figure 14. Lowest $I$ value of some angle sections

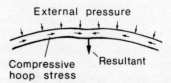

External pressure

Compressive hoop stress    Resultant

Figure 15. Buckling under external pressure

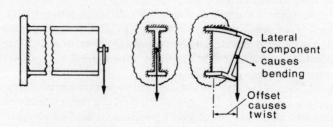

Lateral component causes bending

Offset causes twist

Figure 16. Description of twist–bend buckling; see appendix for analysis

16

the domed shapes used in pressure vessels, design rules are given in the pressure vessel codes[10,11]; these shapes usually depart considerably from the spherical where they blend into other shapes and are not amenable to treatment by this formula because of out-of-line effects.

Another form of failure is the twist—bend or canting failure of beams. It is most readily understood from a cantilever. Figure 16 shows a cantilever fixed to a wall and carrying a load at its free end. If an accidental sideways deflection occurs, it produces a torque about the fixing; the member twists and because the sideways $I$-value is lower than the vertical one, the lateral component can cause appreciable additional bending, further increasing the torque. Buckling instability ensues if the down-load is large enough. (Further details are given in the appendix.) This problem occurs in floor joists; building regulations require the use of stabilising members, sometimes called sprockets, when the span exceeds a certain value. These are of quite light section since they only have to prevent the initial lateral deflection. In engineering situations twist—bend failure can be a problem in crane work, long road-trailers and general handling-equipment design.

The web of an I-beam is similar to the line- and angle-fixed strut of figure 13 but of uncertain width. The main forms of this buckling are shown in figure 17. Figure 17a shows a localised buckle, and figure 17b shows a web panel buckling between supports. In some thin aircraft-spars this buckling was permitted (figure 17c). Clearly this action does not necessarily cause the structure to collapse, since even in the buckled state the structure is equivalent to a framed structure (figure 17d), but stresses and deflections are affected. The calculation of these conditions is found in most books on design, codes for structural steelwork or airframe-design manuals, and it is sufficient here to bring it to the reader's notice qualitatively.

Curved beams have a source of weakness often ignored in textbooks; the tensile and compressive forces are 'out of line', producing a reaction. This can be important in hollow sections and also in channels and I-beams. Consider a curved box-section beam (figure 18). The tensile forces have a downward resultant, the compressive forces an upward one. Near the webs there is no trouble; away from them the section tends to distort so that the flanges do not support their full share of the load.

I-beams and channels are readily stabilised by gussets (figure 19a) but the welds are detrimental under fatigue conditions (see chapter 7). Large hollow beams have internal supporting diaphragms (figure 19b). These need to be strong in buckling, especially if the beam sides (webs) are sloping. The out-of-line effect is particularly striking at corner joints. An unreinforced corner was shown in figure 18; the most common reinforcement is a diagonal plate (figure 19c), suitable for both open and hollow beams. It is, however, a considerable stress-raiser and is sensitive to typical plate flaws (figure 19d)[12]. An alternative is to use substantial side-plates (figure 19e).

Where space is restricted the author suggests using diagonal straps (figure 20). Fuller details may be found in Polak[13]. Basically the force in each flange is found by normal bending theory. A triangle of forces at the joint angle gives the total force along the joint plane and this is allocated partly to the web, partly to the added straps, enabling a suitable cross-sectional area of straps to be decided. The added straps should be overdesigned to allow for imperfect welding. Under static

17

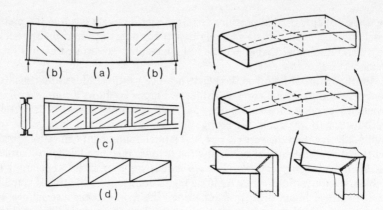

Figure 17. Web buckling     Figure 18. Profile distortion

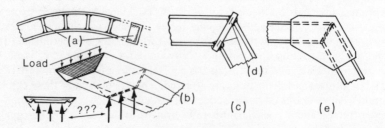

Figure 19. Some reinforcements

Figure 20.

18

loading the reinforced joint shown was almost as strong as the bare beam itself, $2\frac{1}{2}$ times as strong as the unreinforced joint. Aircraft engineers tend to call any such features 'kink-plates'.

An important class of failures results from fatigue. It is well known that materials can eventually fail after many repeated or reversed loadings. For a fuller treatment the author particularly recommends Heywood[14] and/or Forrest[15], supplemented by detailed data such as Peterson's world-wide collection[16] (based mainly on photoelastic work) and various E.S.D.U. issues.

The most common loading is a shaft rotating under steady bending loads so that each point on the surface comes alternately under tension and compression. This will cause failure if the stress exceeds the endurance limit, an ascertainable material property, except for complications connected with stress gradients. Usually the case is also complicated by notch effects, better described as stress-raisers, which magnify the local stress compared with the nominal stress to which a plain member would have been subject. Often these can only be avoided at great expense and it is useful to have some idea of this magnification, expressed as a stress-concentration factor. A few examples are given in figure 21; attention is drawn to the basis of comparison. Where the actual shape is difficult to analyse the local stress is compared with that in a relevant plain member, the effects of shape and of reduced cross-section being lumped together. In other cases a more scientific basis is adopted.

The most common traps in fatigue design are connected with size effect, shrink-fits, corrosion and residual stress. In addition there is the elementary matter of remembering that some calculations give a shear stress whereas the endurance limits quoted for various materials are usually in tension and compression, the endurance limit in shear being perhaps 0.6 of that in tension/compression for many metals.

Even test results cannot always give the whole story. We can take care of size effect by testing full-scale models, we can imitate the corrosive environment and there are many reasons for believing that the frequency of cycling does not invalidate the results, but still we cannot simulate service life which tends to contain long periods at zero load or at steady load, when residual stresses can build up or fade away.

Size effect is related to grain size and crack growth. In theory a sharp notch has infinite stress, yet many sharp-cornered components are used and do not fail, particularly if the component is small. In such cases any incipient cracks soon come into regions where the stress is much lower than at the corner and proceed no further. This property of small components can be most treacherous to the designer if an existing design is scaled up or a big design is 'proved' by a miniature test-piece made of the same material as the eventual part. It is rarely practicable to scale down the grain size, therefore a calculated allowance is made. It is generally believed that in large parts the stress concentration is equal to that found photoelastically or mathematically.

The point is demonstrated in figure 22, adapted from Heywood[14], showing fatigue-test results on specimens of identical geometry but varying in size, cut from identical material. The smallest specimen shows practically no notch effect, being almost as strong as a smooth shaft; the largest was weaker than predicted by theory, showing the difficulty of producing repeatable results in fatigue work. Such

19

a specimen in effect only tests a very small region of material at the shoulder fillet so that variations in the material tend to have large effects.

Some of the older fatigue-test data should be treated with caution in case they are derived from very small specimens subject to this misleading effect. Another possible error source is inertia of the testing machine; for safety the stresses should be measured by strain gauges on the specimen itself.

Shrink-fits to assemble large components rigidly are used in railway wheels and axles, turbine rotors and marine-engine crankshafts, to mention just some better-known examples. Generally the interference is calculated so as to ensure a contact pressure of such intensity that the frictional grip is not disturbed by the working stresses in the finished assembly; often the mating surfaces are partly relieved to achieve this. The reason is that if slippage should occur, fretting corrosion may start.

Some shrink-fits are assembled by freezing the inner part in liquid carbon dioxide or liquid nitrogen. As the outer member comes into contact with the cold inner one, the surface may cool down to the brittle fracture region (see chapter 6) while building up tensile hoop-stress (bursting stress). It may be advisable to pre-heat the outer member to a temperature at least as far above ambient as the inner one is below ambient.

If a shrink-fit is tight enough to prevent slippage, the assembled article behaves as one body, often with very severe stress-raisers. This was reported long ago by Kuhnel[17] whose work was reviewed and augmented by Peterson and Wahl[18]. This important effect has not featured enough in education, with the result that a later generation of engineers has run into the same problems. Coyle and Watson report failures on these lines[19].

The conclusions are simply that an assembled component must be regarded as solid, and thus may have some very severe corners or notches. The effect of deep notches can be taken from Peterson[16]. The point is illustrated in figure 23 which also shows the relief used to improve grip.

Fretting occurs where local slippage disrupts the oxide film on which most metals depend for protection. It has been investigated particularly in aluminium alloy lugs secured by steel pins, formerly a common aircraft detail. Instead of the theoretical stress concentration of $2\frac{1}{2}$ to 3, life tests point to a factor of 10 to 20. The shape of the hole distorts under load and local rubbing against the pin takes place, aggravated by the discrepancy in Young's modulus between the materials.

Several remedies have been found, each with some disadvantages: a large clearance, an interference fit, a steel bush pressed in, or a chemical treatment to separate and lubricate the surfaces. The best choice depends on individual circumstances such as frequency of dismantling, importance of firm location, etc.

Corrosion in a fatigue situation is always harmful, and no doubt also connected with breaking up of the oxide film. Even corrosion-resistant materials show lower permissible stress for a given life in steam or water (especially salt water), etc., as compared with clean dry laboratory air. The most highly stressed points tend to corrode most (for further discussion of stress corrosion see chapter 6).

Perhaps the most important effect is that of residual stress. Imposing residual compressive stress greatly improves fatigue resistance, often doubling the permissible nominal stress level[20]. This can be done by treating the part with a polished hard steel roller pressed against the rotating surface (a long-established railway

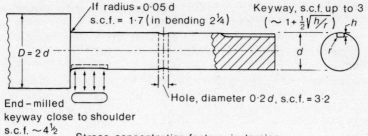

If radius = 0·05 d
s.c.f. = 1·7 (in bending 2¼)

$D = 2d$

End – milled
keyway close to shoulder
s.c.f. ~4½

Keyway, s.c.f. up to 3
$( \sim 1 + \frac{1}{2} \sqrt{h/r} )$

Hole, diameter 0·2 d, s.c.f. = 3·2

Stress concentration factors in torsion.

Shear stress = torque × s.c.f. × $\dfrac{16}{\pi d^3}$

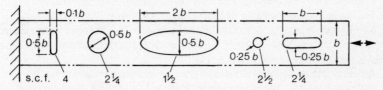

| s.c.f. | 4 | 2¼ | 1½ | 2½ | 2¼ |

Holes in flat bar, tensile s.c.f. based on load ÷ net cross section

Figure 21.

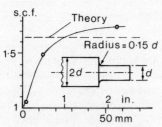

Figure 22. Effect of size
on endurance test results

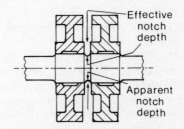

Figure 23. Stress-raisers in
shrink-fitted assemblies

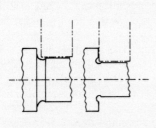

Figure 24.

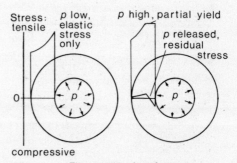

Figure 25. Autofrettage

21

practice) or shot-peening which affects the surface finish and is appropriate for springs. Some case-hardening treatments, notably nitriding, also induce compressive stress. These stresses normally arouse a balancing tensile stress below the surface which is generally harmless but should not be forgotten.

Tensile stresses in a surface are generally harmful. One source of tensile stresses is overheating due to grinding; it is often advisable to design components for grinding so that shoulder radii are machined with a slight undercut, as shown in figure 24, so that grinding damage does not coincide with the geometric high-stress region. A slight run-out also helps in the grinding operation itself.

Perhaps the most common source of tensile residual stress is weld shrinkage. Unless stress-relieved, a weld almost certainly has very high residual stresses.

Another likely source, not yet fully investigated, is decarburisation of carbon steels. During hot-working or heat-treatment, carbon is likely to be lost from the surface layers. When the opposite happens, in case-hardening, a compressive surface stress is produced. Thus it seems likely that where carbon atoms are oxidised away a tensile stress may remain (see p. 44).

An early use of deliberate residual stress was autofrettage of gun-barrels. The reader may have met the elastic stress distribution in thick cylinders under internal pressure (figure 25). In autofretting a cylinder is pressurised so highly that the inner layers of the gun are stressed well beyond the elastic range of the material and some regions deform plastically. On relaxing, the part of the gun-barrel which had not yielded tries to return to its previous size but is stopped by the inner layers which have grown too big. Thus in the final barrel the inner layers are in compression. The compressive stress provides resistance to cracking due to firing pressure and thermal cycling.

Autofretting occurs in a less definite manner in any component containing stress-raisers when it is first loaded. Many springs are bumped or scragged to their maximum deflexion several times as part of production. This allows yielding to occur at high-stress points such as at the inner surface of helical springs or at discontinuities in the material. In subsequent service the stress can never exceed yield point provided that the spring never receives reverse load; therefore springs can be designed to operate at quite high stresses (see Wahl[21]).

Structures and pressure vessels normally receive a proof or test load in excess of the stated service load. Any discontinuities will either crack or receive a residual stress opposed to the working stress; provided no cracks are started, the working stresses are thereby lowered wherever yielding had occurred. In pressure vessels reverse loading is rare, arising mainly from thermal expansion on start-up, cool-down, or temporary flow stoppage, for example, when steam flow has to stop suddenly and the steam lines cool down, or in vessels with a heating jacket fed by steam.

In cranes and bridges carrying moving loads the shear members may be subject to forces changing from tension to compression; these members may actually be weakened by the overload test. Accordingly a special stressing rule is made in some design codes, the force for design purposes being the maximum force plus half the maximum reverse force in the member concerned.

One example of a lack of understanding of the residual stress effect occured in the aircraft industry, with drastic consequences. A high-altitude airliner with pressurised cabin was tested in prototype like a pressure vessel, in fact more so, to

22

twice the intended working pressure whereas $1\frac{1}{2}$ times is common practice in pressure vessels. The same prototype was then fatigue-tested to above service pressure, being pumped up and released over 16 000 times, many more loadings than would ever occur in service. However, the craft released for service were pressure-tested to only $1\frac{1}{3}$ times working pressure. After two aircraft failed in flight, fatigue at a stress-raiser was suspected, particularly in view of reports of cracks, repaired according to current practice. Another machine, as yet unflown, was pressurised repeatedly and failed after 3000 cycles[22]. In the context of this chapter it is not hard to see the reason. The prototype was autofretted so well that its discontinuities were thoroughly pre-stressed in the favourable direction whereas craft in service did not have such beneficial treatment. There is further evidence that a test to $1\frac{1}{4}$ or $1\frac{1}{3}$ times working stress does relatively little to protect a vessel against fatigue, whereas a higher loading, in this instance accidental and of unknown magnitude, is very beneficial (see Taylor[23]).

In stress calculations it is tacitly assumed that we know the magnitude of the load. In pressure vessels and boilers the peak loading by pressure is highly predictable, any attempt at overpressure causing safety-valves or bursting-discs to blow. Even explosive conditions are to some degree predictable, only stresses due to improper support and thermal expansion being doubtful. In buildings and civil-engineering structures the levels of stress and the loads to be assumed, hopefully in excess of any possible service load, are laid down by design codes to which one has to conform — in some cases by law, in others to satisfy the insurance company.

In aircraft also the loads and stress levels are laid down by safety authorities, but owing to the obvious need for weight limitation it is recognised that over-stress may occur due to abnormal air gusts or hard landings. In view of this the detail design is arranged as far as possible so that weak points come in places readily accessible for inspection and repair but with plenty of redundancy so that the danger is not immediate.

In cars the peak loads are quite unknown, being due to impacts with kerbs, to potholes and to emergency manoeuvres which are much more severe than normal duties. Here the designer has to rely on past experience, designing new details to be as strong as satisfactory old ones. Machines liable to jamming should incorporate a release coupling, either of proprietary make or designed-in; at least a shear pin should be fitted.

A difficulty occurs when the loading is mainly steady, with a small fluctuation added. For a steady load we tend to ignore stress concentrations, allowing local plastic flow to smooth out the worst. In springs and in many other cases such initial settlement is acceptable, so that the stress-concentration factor need be applied only to the fluctuating part of the stress. In other cases it is safer to take the attitude that using the stress-concentration factor is simply a convenient method of finding the real stress.

The procedures for relating the material properties such as endurance limit, U.T.S., etc., to any particular loading range are given in most design books, for example Phelan[24]. They are based either on the lowest and highest stresses or on the mean value and superimposed amplitude or range. The reader is advised to look out for confusion between the terms range, full range, ± range, half-range and amplitude.

Heywood[14] gives data on combined bending and torsion fatigue. Unfortunately the most common case, fatigue bending under steady torque, has not been investigated very fully. Further discussion of design stresses will be found on p. 62.

In some cases, especially in aircraft, economy or weight limits lead to designing for a limited life. The $S{-}N$ curve describes the number of loadings ($N$) at any given stress ($S$) before failure is likely; in practice we are faced with a whole spectrum of load cycles. They are usually taken as cumulative; for example, if the material can survive 10 000 cycles to peak stress $A$ *or* 1 000 000 cycles to peak stress $B$ it can also survive 5000 cycles to stress $A$ *plus* 500 000 cycles to stress $B$, etc.; at each level of stress an appropriate fraction of the life is used up (Miner's law, or more properly the Palmgren–Miner hypothesis).

# 4

# Stable and Unstable Systems

Most practical systems are required to be stable, at least dynamically. Consider a bicycle; it is statically unstable, yet a small child can learn to ride it. How is intrinsic stability obtained? There is a popular notion that the head angle confers stability, yet early bicycles had upright steering heads. A little thought shows that sloping heads actually tend to de-stabilise. Under gravity the load and earth try to come together. Relative to the bicycle, the ground point P as defined in figure 26 tries to come upwards from the lowest position at straight ahead, which it does by rotating the steering. This is easily confirmed in practice; the equilibrium point comes at a steering angle of 60 to 80° from straight. The stability comes from the trail. If the bicycle leans sideways, the ground-reaction has a lateral component steering the wheel towards the leaning side until centrifugal force restores balance. If the trail accidentally becomes negative, riding hands-off becomes very difficult. Some small-wheeled bicycles were designed on the basis of head angle and swept-forward forks, leaving too little trail.

Cars need stability in cornering such that centrigual force does not swing the tail out and sharpen the rate of turn (oversteer). This is obtained in the first place by putting the centre of gravity before the midpoint of the wheel base; since tyre side-drift is mainly proportional to side-force, the front drifts out more than the rear, which is stable. Note that side-drift is also affected by down-force, driving or braking torque, local down-load, wheel-camber and inflation pressure. In the old days it was necessary to increase the rear-tyre pressure when the rear was heavily laden in order to stiffen the rear tyres against side-drift. Modern cars handle in the same way whatever the load, within limits. How is this achieved?

The rear wheels are suspended in such a way that the side-force itself, or the consequent body-roll, steers the rear wheels just enough to compensate for the change in side-drift. Both these ideas are described in the next chapter, not as exercises in the subtleties of automobile engineering but as linkage design examples. The roll-steer principle is almost as old as the motorcar itself since semi-elliptic ('cart') springs tend to give this effect automatically to a slight extent.

Aircraft need inherent stability in several respects. For example, consider a sudden increase of incidence: this gives the wing extra lift, often also bringing the centre of pressure further forward. The tailplane should gain lift at a greater rate than the wing, otherwise the aircraft could nose-up rapidly before the pilot has time to correct, overstressing the wing or possibly causing stall. It is particularly important not to get the tail surfaces into the broken airflow from a stalled wing.

To ensure stability, the down-load on the tail is kept low or even negative, so that it has plenty of lift in hand and a steep lift-curve. Since the using-up of fuel can displace the centre of gravity very substantially, the disposition of tanks is planned to suit and the sequence for usage of fuel from the various tanks is carefully laid down.

To change course aircraft are made to lean over, or bank, like a bicycle so that the lift force gets an inward component. If bank is insufficient (or excessive) side-slip takes place, and for inherent stability the side-slip should automatically correct the banking. This can be achieved either by dihedral (sloping the wings upwards from root to tip) or at least by putting the lateral centre of pressure above the centre of gravity. The lateral centre of pressure is somewhere near the centroid of the side-view outline so that tall tail-fins provide some high-up area very conveniently; low-slung engines keep the centre of gravity low.

An interesting problem arose from the method of obtaining bank in early aircraft. To obtain bank, one wing must be lifted while the other is dropped. In some cases this was achieved by warping the wings; with more modern rigid wings, ailerons are used. The trouble arose from the increase of drag when trying to lift a wing, which was sometimes enough to steer the craft in the opposite direction to that required. Once this problem was understood it was easily remedied by differential aileron control, making large upward deflections at the inner wing-tip but much smaller downward deflections at the outer one. One possible linkage layout for this is shown in the next chapter.

In stability considerations we can generalise thus: a disturbance should produce a correcting response, preferably without involving control apparatus. If the response is too weak it may be incapable of correcting a strong disturbance. If it is too strong or too persistent it will produce overshoot and probably oscillation. This book will not go into stability criteria, but merely discuss the effect qualitatively: some oscillations die out, while in other cases there may be a power input if there is interaction between the prime cause and the motion, especially if there is a certain phase-relation between them. This is called positive feedback.

Clocks and electric bells supply this power input in a deliberate, organised manner by providing a driving force cyclically, in phase with the motion. Musical instruments also produce sustained oscillations; they take power from the player and use natural phenomena to ensure the phasing. The phenomenon used by stringed instruments is speed-dependent friction, or stick—slip to use an over-simplified but descriptive term. Wind instruments use vortex shedding; this also occurs very commonly in engineering — for example, suspension bridges, galloping power lines and swaying chimneys. To describe the phenomenon in detail would take us into aerodynamics; briefly we can say that when a cylinder or other body oscillates, the apparent (relative) wind direction tends to give improved flow over the advancing surface, thus giving it a lift force at every swing.

Some examples of remedies are shown in figure 27; suspension bridges which

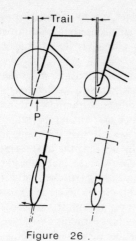

Trail

P

Figure 26.

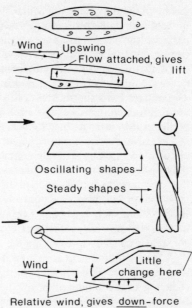

Wind  Upswing
Flow attached, gives
lift

Oscillating shapes
Steady shapes

Wind
Little
change here

Relative wind, gives down-force
during upswing, stops oscillation

Figure 27.

Figure 28.
Relief valve with
reaction surface

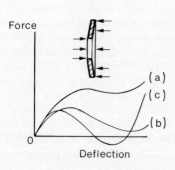

Force

(a)
(c)
(b)

0

Deflection

Figure 29.

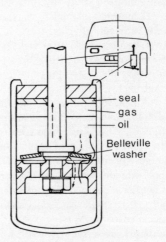

seal
gas
oil

Belleville
washer

Figure 30. Vehicle damper,
emulsion type, simplified

27

can be particularly unstable when close to the water, can be stopped from oscillating by adopting stable profiles as shown (Wardlaw[25] shows limited amounts of flow patterns; the explanations are by the author). On chimneys, helical ridges cause the eddies to cancel each other's effects.

Oscillation due to over-controlling is discussed next. Consider a furnace: the doors are opened to insert or remove an object, producing a sudden heat loss. The furnace has a thermostat consisting of a pyrometer with high- and low-temperature contacts which switch the power on or off. The temperature thus oscillates over a range; inevitable lags in the system mean that this range may well be excessive. A simple improvement is obtained by dividing the heaters into two circuits, operated by separate contacts on the thermostat, so that a small change only operates the first, closely set contacts, the rest being switched only if this does not provide enough control. In most cases this idea is taken to its logical conclusion of proportional control, responding in proportion to the 'error'. With fluid systems such as gas or oil firing this is not unduly difficult. With electric heating the control is sometimes arranged by a deliberate on—off cycling at constant frequency but varying the on—off (mark—space) ratio.

A further refinement is to take the medicine before the headache. A sudden load change can be detected by noting the speed with which the temperature is changing and switching on or off prematurely to suit; this is error-rate sensing. It is not unduly difficult to sense error rate; for instance an electrical signal is easily differentiated with respect to time.

This system tends to take too much notice of a momentary high demand and also tends to run too low by an amount which depends on the load. If we time-average this deficiency by integrating the error signal and use the answer to adjust the original set-point accordingly we can cure both these troubles. This facility is built into three-term controllers with a proportional band, a differential action for error rate and integral action for smoothing and reducing the 'offset'. It is emphasised that such a system cannot be designed and made stable by qualitative common-sense alone; a study of control theory is essential.

Sometimes instability is useful. A steam safety-valve should act by giving a good long blow and then shutting. If it has progressive action it would dribble at slight overpressures and suffer wear from the scouring of high-speed steam over its face. By providing extra lifting surface which becomes effective only when flow has started, it lifts at the design pressure but does not re-seat until the pressure drops to a per-ceptibly lower pressure, perhaps 90 per cent of rated pressure in some cases (figure 28).

A Belleville washer incorporates all the elements of an unstable system in one simple component, a conical disc with a central hole. When loaded as shown in figure 29, the outer edge tries to lengthen, the inner one shortens; this results in an opposed system of forces. As the disc flattens, the planes of action of these forces approach and can eventually cross over. By choosing the ratios of diameter, hole size and wall thickness, various load—deflection curves can be obtained. Case (a) in figure 29 with its nearly constant-force region is used in car clutches and also as overload relief valve acting both ways in the de Carbon damper widely used in car suspensions (figure 30). Case (b) would make a bell-push or pressure-switch. Case (c) snaps right over and would make a maximum force indicator or an on—off switch. Almen and Laszlo[26] give extensive details about loads and stresses in Belleville washers or disc-springs.

28

# 5

# Some Motions

A prime example of straight-line motion is the lathe bed whose straightness, if parallel with the axis of rotation, generates accurate cylinders. It is usually guided by three surfaces arranged in V-flat formation as in figure 31a, giving unique kinematic location so long as the loads are downwards. Where this is not certain, slides are often of dove-tail shape, with three fixed surfaces, the fourth being adjustable, as in figure 31b. If we provide four rigid surfaces, we are over-located as in figure 31c, which shows in exaggerated form the consequences of slight error such as may occur due to distortion, thermal expansion, etc. The three-surface location is free from this trouble; when the V-slide is symmetrical and wear is even on both sides, the wear causes no slackness and no horizontal error. Note that the cutter position is arranged so that wear of the slides or cutter deflection cause only small, second-order errors in the workpiece.

Locating for thermal expansion, for example, in steam turbines, can be arranged to keep the axis (or any other chosen point) unchanged despite temperature changes, by arranging the slides to converge on the fixed point, as in figure 32 adapted from Church[27]. Uniform thermal expansion produces a change of size but not of shape; therefore we can expect angles to remain the same and radial lines to continue to radiate from the same centre.

Ordnance Survey triangulation points seen on many hill-tops are instances of three-point kinematic locations which are unaffected by temperature. Three horizontal V-grooves receive the conical or ball-ended feet of the theodolite (figure 33). The feet need not have any particular spacing so long as they are symmetrical about the axis.

Straight-line motion can be transferred and magnified by linkages. An example is the lazytongs lamp- or tool-holder for locating spot-lights or heavy tools such as spot-welders anywhere above a working area. The vertical guide on the wall-bracket ensures horizontal motion at the hooks so that gravity plays no part and the load stays put. Figure 34 shows this and also a curved lazytong motion which is produced by making the links slightly unequal. This latter example needs no guide. Use of a guide-bar involves friction and wear, therefore one should be aware of straight-line motions produced by linkage only. Figure 35 shows Peaucellier's method

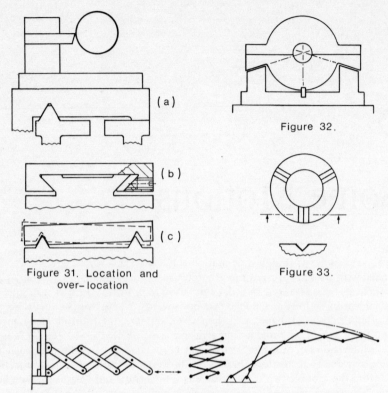

(a)

(b)

(c)

Figure 31. Location and over-location

Figure 32.

Figure 33.

Figure 34. Straight-line and curved lazytong motions.

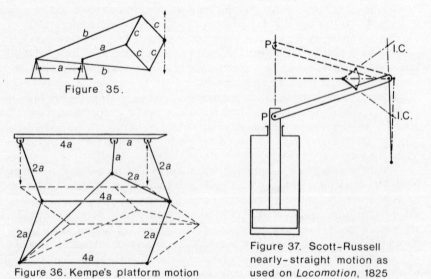

Figure 35.

Figure 36. Kempe's platform motion

Figure 37. Scott-Russell nearly-straight motion as used on *Locomotion*, 1825

30

producing true straight-line motion of a point. An extension of this is Kempe's motion for a bar or platform with true vertical movement while the bar remains horizontal (figure 36). These figures are adapted from Beggs[28]; another, more recent book on motions is by Tuttle[29].

Linkages are best studied graphically, on a large sheet of graph paper overlaid with tracing paper. Once a basic grasp has been obtained the process can be programmed on a computer, especially if a graphic facility is available.

Approximate straight-line motions are of practical interest, especially in materials-handling machinery. One, a Scott–Russell motion, guided the piston rod on Stephenson's *Locomotion*, of 1825 (figure 37). The links must be in certain proportions so that the instantaneous centre (I.C.) keeps level with point P. The two basic Watt linkages are shown in figure 38. The instantaneous centre still tends to follow point P.

An indirect use of a Watt linkage produces the vertical motion of pantograph current-collectors for electric railways. These require vertical, not angular, motion, so that trains can go forwards or backwards without trouble arising from friction-induced bounce at the contacts; the moving parts should have low friction and low inertia, making a linkage the natural choice (figure 39). These collectors are not pantographs in the classical sense; a true pantograph uses parallelogram linkages to enlarge or reduce the motion of a point. In Figure 40 this is applied to an engraving machine for cutting letters of any size from a single master set. Simple scales employ the parallelogram to give parallel motion of the load pan and weight pan so that weighing action is not affected by detail positioning of load or weights (figure 41). Larger platform scales achieve parallel motion by pairs of links of unequal length but identical leverage ratio.

A car contains several instructive linkages. A good steering layout for long tyre-life ensures minimum departure from true rolling. Kinematically a wheel (cylinder) can roll only in a straight line and a cone is needed for curves. Indeed, a wide cylinder has a substantial straightening-up tendency, noticeable in garden rollers. This is called self-aligning torque in the automobile business; some aircraft writers use this term for what would better be called castor-effect; perhaps the toroidal aircraft tyre has less self-aligning torque than a car tyre. For minimum scrub all wheels should be tangential to the curved path, all axles converging on a common point. This is called Ackermann steering, and was invented before cars (figure 42). In racing cars, where large drift angles are usual and tyres do not last very long, this geometry is less relevant, the steering angles being such that the actual drift angle of each tyre bears some relation to the down-force at that point.

The steering geometry of a particular car shown in figure 43a has large amounts of scrub at high lock-angles. A better layout shown in figure 43b is capable of equally tight circles but with much better geometry. The scheme in figure 43c has even better geometry but more restricted extreme lock. The angle $\theta$ between links should preferably not exceed 150 to 160° or it may become difficult to return to centre.

The next example relates to preventing interaction between steering and vertical motion of the suspension. (Some cars have deliberate interaction giving toe-out on upward wheel displacements (bump) and toe-in on downward ones ('rebound' in automotive jargon) supposedly to balance out camber thrust. The author feels that this is likely to be detrimental to road-holding and tyre life.) Consider the popular

31

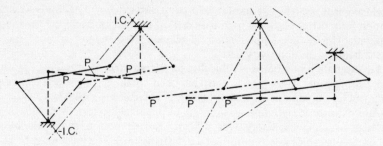

Figure 38. Watt linkages for nearly-straight motion

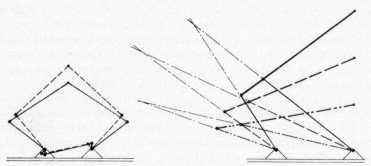

Figure 39. 'Pantograph' current collector layouts

$$\frac{a}{b} = \frac{c}{d} = k\,(\text{adjustable}) = \frac{TP}{PC}$$

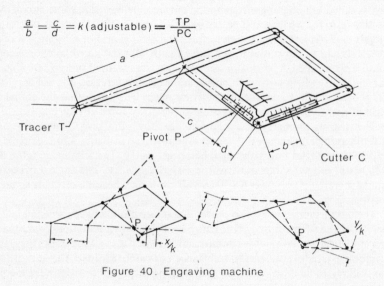

Tracer T

Pivot P

Cutter C

Figure 40. Engraving machine

32

wishbone linkage. The swinging link of the steering mechanism must move in unison with that suspension link which lies in the same working plane. If this is not convenient we merely have to put in an imaginary link which has the same instantaneous centre as the suspension linkage. The working plane may be horizontal, as implied in figure 44, or inclined. Inclined working planes are sometimes used to reduce impact forces or to reduce nose-dive on braking, as mentioned below.

If the linkage in side-view gives the front and rear ground points a motion perpendicular to a line from the centre of gravity we should get no pitching due to either braking or acceleration: this is called anti-dive and anti-squat (figure 45). Usually partial anti-dive or anti-squat is found sufficient. As mentioned in chapter 4, cars and trucks tend to oversteer when the back is heavily laden, the increased side-force at the rear gives greater tyre drift, and the back tends to swing out more than the driver intends or expects. This can be compensated in two ways. Direct use of the side-force is shown in figure 46, omitting the main springing for clarity. The axle is located by inclined rubber shear pads or by angled links. It is allowed to deflect slightly sideways (which is detrimental to stability in extreme cases) and as it moves it skews slightly, pointing into the bend and thus compensating for the outward drift in proportion to the side-force.

Figure 47 shows how this compensation can be obtained without allowing axle side-movement by using body-roll which is also proportional to side-force although it takes a little time to develop. Omitting a great deal of detail essential for the vehicle but not essential for this explanation, the rear axle is located by arms A. At light loads these arms slope downwards and as the car rolls (heels) outwards the right arm gets steeper and drags the right wheel forward. Similarly the left wheel is pushed aft. The axle skews and steers the back outwards, adding to the tyre drift which of course is small at light loads. At medium load the arms are horizontal and action is neutral; at high rear loads (in both senses) the arms slope up. This time body roll pushes the right wheel aft while the left goes forward, skewing the axle into the bend. The rear is now steered inwards to compensate for the greater side-drift. (A high-up load tends to cause more drift but fortunately also gives more roll and hence more compensation.) This effect is not confined to rigid axles; it can be designed into independent suspensions too[30].

Another instructive linkage problem arose when car engines were first rubber-mounted. The engine leans over on its mountings according to the torque exerted on the output shaft. What effect does this have on the operating controls? Imagine the car being put in gear and the clutch beginning to engage. The engine begins to exert torque and by reaction leans over on its mountings; if this lean causes the clutch to bite more at the same pedal position, judder is likely. If the links are angled so that the leaning action eases off the clutch, action tends to be smoother (figure 48). Similar arguments apply to the throttle linkage; increasing torque should cause the engine lean to ease off the throttle. Cable and hydraulic controls are neutral in this respect.

Students sometimes feel that the study of linkages is empirical, without any general logic; the following principle is therefore put forward. Consider a rotating shaft, with an arm A connected to link L. When A and L are at right angles, the output speed is greatest and the force is least. At other angles the effect is as for a shorter arm whose length is equal to the offset. When the offset is zero, we get zero speed but infinite force, called toggle action (figure 49).

33

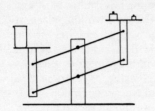

Figure 41.

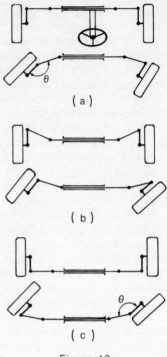

( a )

( b )

( c )

Figure 43.

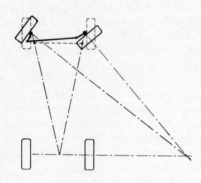

Figure 42.
Ackermann geometry

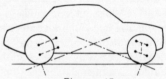

Figure 45.
Anti – dive and anti – squat

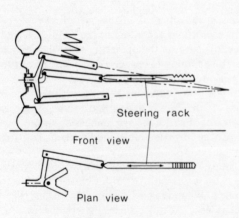

Steering rack

Front view

Plan view

Figure 44.

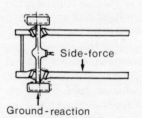

Side-force

Ground - reaction

Figure 46.

34

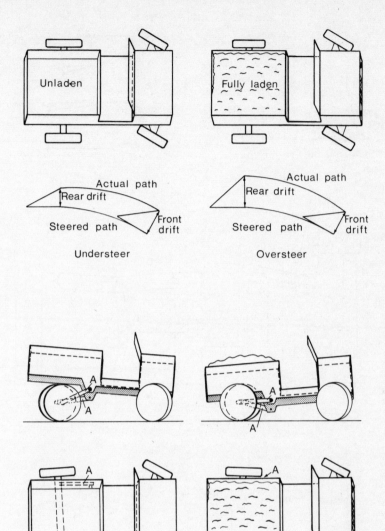

Understeer                    Oversteer

Compensation produces constant actual path at all loads.

Figure 47.

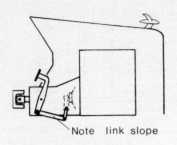

Note link slope

Figure 48.

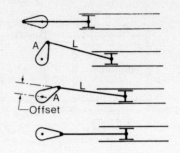

Figure 49.

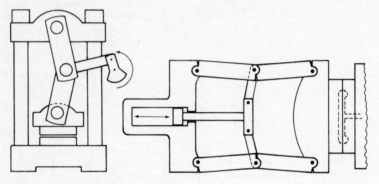

Figure 50. Toggle action

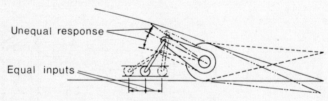

Unequal response

Equal inputs

Figure 51.

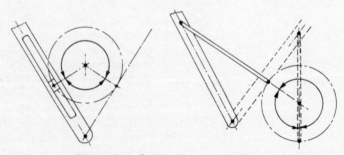

Figure 52. Quick-return motions.

This action is sometimes duplicated, as in the toggle (coining) press and the die-holding arrangements of plastic moulding and diecasting machines. In these an extra feature can be provided, namely abutments on the links to take the major load off the pins (figure 50).

The idea of varying the offset is used in the mechanism shown in figure 51. This is purely an illustration but could be a way of achieving the differential aileron control discussed in chapter 4, practical versions of which are more likely to form part of a three-dimensional layout harder to depict clearly.

Indirectly the same principle is found in quick-return motions such as the well-known Whitworth slotted-link type. It is less well known that a similar effect can be obtained by links alone, without slider (figure 52).

The most versatile and straightforward mechanism for producing a complex motion is undoubtedly the cam. Though we tend to think of engines as the main users of cams, a home sewing-machine contains a far wider range of cam types and is a much more accessible subject for study. In industry, automatic lathes and much other process machinery would be vastly more complicated but for the use of cams.

The most general form is shown in figure 53a. The required motion in this example consists of a slow approach to take up any initial clearance without undue noise, then an upward acceleration to a steady rise speed. Next comes a rapid deceleration where the return springs have to keep the follower from bouncing off, leading to a dwell at maximum lift. The return motion is symmetrical with the rise. The springs supply the force for the initial downward acceleration, with some to spare to maintain contact and prevent the roller from skidding. We require this motion at the plunger P, produced by some convenient lever system with a smooth hard boss bearing on the plunger and a roller bearing on the cam itself.

The required time-cycle which often contains a substantial dwell at base level enables us to decide the camshaft speed; of this the active cycle occupies some portion, divided into equal angle intervals. The desired motion is divided into the same number of equal time intervals. The lever and roller position for each instant are set out and transferred round to the appropriate angle. The successive roller positions give the envelope of the required cam shape. The reader is advised to follow this out on the drawing-board, starting with an arbitrary system and motion of his own choosing.

If we think of the motion as a perturbation imposed on a basic circular path, we see that there must be some lower limit on the size of the basic path. This is illustrated in figures 53b, c and d. The cam in figure 53b has the same base circle and motion as in figure 53a but with a flat follower or tappet. This is much simpler to set out; the rise at each angle gives the follower position directly, enabling us to draw the cam and at the same time to see how large the flat follower has to be to maintain contact at the tangent points. In figure 53c the same motion is plotted on to a greatly reduced base circle and the profile has become unattainable. This is known as interference and particularly affects cams with flat followers. Interference can be prevented by substituting a convex follower, be it a roller or a rounded slider (figure 53d).

Automobile cams are well lubricated so that rollers are rarely needed, the cheaper flat tappet being adequate. To distribute the wear evenly, tappets are encouraged to rotate by offsetting the cam to one side, approximately as in figure 54, or alternatively by slightly tapering the cam to give more surface pressure at one edge.

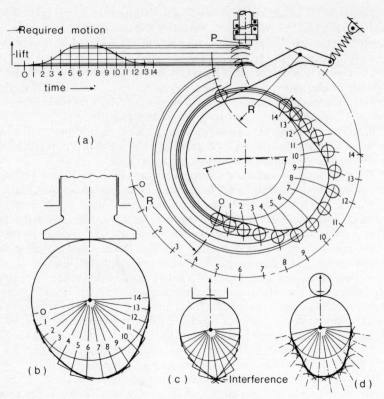

Required motion

lift

time

P

R

(a)

R

(b)

(c)   Interference

(d)

Figure 53. Cam profile design

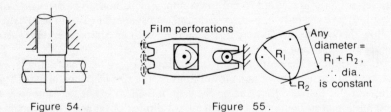

Figure 54.

Film perforations

Any
diameter =
$R_1 + R_2$,
∴ dia.
is constant

$R_1$

$R_2$

Figure 55.

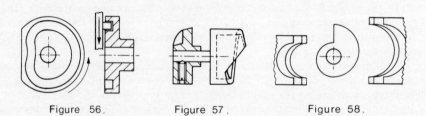

Figure 56.          Figure 57.          Figure 58.

38

A cam profile is not entirely arbitrary; we have to restrict the accelerations so as to limit the forces on the cam and to keep the return-spring requirements within bounds. Traditionally cam profiles were aimed at constant acceleration and deceleration, giving parabolic curves of travel against time. Lately it has been realised that the return-springs exert somewhat greater force when more compressed so that decelerations and accelerations at the top of the travel can be higher than those in the lower parts, making that part of the motion sinusoidal.

Some cams do not need return-springs, for example, the positive cam used in some cine cameras and projectors for film transport (figure 55). The cam is of constant diameter which just fits the dimensions of a square window and gives it a roughly circular orbit three times per revolution. By virtue of the peg and slot, the teeth at the film end have the vertical component exaggerated while the horizontal one is almost unchanged. The teeth thus have an elliptical motion very suitable for engaging and moving the film, while the light-beam is cut off by a rotating shutter blade. While the picture is projected or received, the teeth disengage and return for the next bite. This is best demonstrated with a large cardboard model. High-speed cameras use constant film speed and a cam-operated rocking prism to shift the image in phase with the film. Ultra-fast cameras require fully rotating optics.

A track-type positive cam is used to move the needle in some sewing-machines by way of a roller fixed to a vertical plunger; figure 56 shows a simplified version. An axial equivalent of this is used in one type of car steering-box.

Simple axial cams are used to set the temperature or time in electric irons and toasters (figure 57). Snail cams as in figure 58 are useful for clamping or adjustment. Often they need not have an entirely regular rise-rate but can consist of two offset semicircle shapes which makes for ease of manufacture by standard concave milling cutters.

# 6

# Materials in Outline

The designer must obviously know a great deal about materials, their properties and problems. The maker of the object also should be familiar with these; to call for materials new to the maker can produce problems of supply, identification, treatment and storage. So whenever possible the designer should check up on this point and possibly change one or the other, material or maker.

The chief properties of interest to the designer are strength, density, rigidity, toughness, cost, corrosion resistance, thermal expansion and thermal conductivity, maximum service temperature and electrical properties.

The data quoted have been extracted mainly from suppliers' literature, British Standard specifications, Smithells[31], Kaye and Laby[32] and Comrie[33].

The strongest materials per unit weight, at ambient temperature, are those with molecular alignment: boron, carbon and glass fibres, stretched polymer fibres and sheet (Melinex, Mylar), steel wire suitably heat-treated, with titanium, aluminium and magnesium alloys close behind, followed by alloy steels. The fibrous substances show their strength in tension like string; in bending or compression they need to be connected, usually by plastics, otherwise they would bend like string. Unfortunately the elastic moduli of the bonding plastics are much lower than those of carbon or glass fibres, resulting in danger of localised buckling and internal cracking. Where high deflections are permissible, strong polymer fibres bonded with plastic are coming into use. In theory, these should be relatively free from this problem.

Rigidity is somewhat related to melting point and to density, a relationship no doubt originating in the atomic structure. Tungsten and its carbide are over twice as rigid as steel, with chromium and carbon fibre in between. Copper and zinc, magnesium and titanium support this rough rule. Young's modulus is appreciably temperature-dependent. Strength is related to crystalline slip-planes and defects rather than to fundamental structures so it should not surprise us that very similar rigidity is found in a soft pure metal and the same metal strengthened by work-hardening or by alloying.

Toughness is a property the designer likes to rely on for withstanding unexpected structure overloads; ductility is useful in overcoming local, perhaps incalculable overstressing. Components, like rules, are often better bent than broken. There are exceptions to this, notably machine tools and measuring equipment which if distorted could result in defective work. This is one reason why machine tools are largely made of brittle grey flake cast-iron and one firm made its inspection gauges of glass.

To assess toughness one can look at tensile elongation values and also impact values, that is, the energy required to break a standard specimen in a standardised way. An intermediate assessment is obtained by fracture toughness tests which involve fatigue-loading a notched specimen. Gray[34] and Kalderon[35] indicate that lack of impact strength should not be ignored even when the service is impact-free. Low-temperature steam-turbine discs failed by stress-corrosion associated with temper-brittleness (revealed by impact tests), although elongation and fracture toughness both appeared satisfactory. The remarkable regularity of the cracks suggests some additional factors, perhaps excessive cooling-rates.

Cost per unit mass is not very helpful unless we are looking for ballast or counterweights, where water, sand, concrete, steel and cast iron form a rough merit order. For design purposes more elaborate criteria can be thought up, such as how to cover an area. The cheapest covering materials for large areas are (in ascending order of cost) plastic film, wire netting, paper, pulp board, steel, plywood, asbestos board, aluminium and glass. For smaller components the order of cheapness of covering materials is likely to be plastic, steel, aluminium, diecast zinc and cast iron, depending on shape, thickness and required finish. If strength or stiffness are particularly outstanding requirements the cost order for small components tends to change to steel, cast iron, aluminium alloy, zinc alloy and glass-reinforced plastic.

In future we may expect aluminium to move nearer the top on all fronts since it is made electrolytically, often adjacent to sources of water-power. The recently discovered zinc—aluminium—magnesium alloys, which are of excellent forming ability and high strength, may well eventually top the list, especially if they should prove to have good salvage (scrap) value.

It is well known that iron and steel rust readily and progressively because they do not form a self-protecting film. However, most non-ferrous metals have a fair resistance to atmospheric corrosion, some even to sea-water. Even magnesium, suitably alloyed, can resist atmospheric corrosion though usually a chemical passivating treatment is applied.

A rapid form of corrosion is that due to galvanic action. Dissimilar metals in contact or metal in contact with carbon have an electrolytic potential in the presence of water. Current flows locally or through the rest of the machine and the more electronegative material dissolves. Impurities can serve as the other metal and it is found that very pure metals corrode much less than those with impurities or alloying additions *in many cases*, while some alloys are more resistant than their separate constituents.

Even within a single component local cells are set up due to stress which render the most highly stressed part electronegative to the rest and thus encourage preferential corrosion where it is least welcome. Local cells can also be caused by differences in oxygen content of the moisture film in crevices. The head of the crevice corrodes preferentially so that corrosion tends to creep underneath a coat of paint, as motorists know only too well.

41

Thermal expansion can be troublesome. Many plastics, unless heavily 'filled' with glass, etc., have very high thermal expansions, ten times that of steel, whereas borosilicate glass and fused silica have very low expansions. Invar, an iron-nickel alloy, has very low expansion over a wide but limited temperature band. Its use hitherto has been confined to instruments but is likely to extend into machine-tools, especially tool shanks and holders whose thermal changes affect product size. Elinvar has elastic properties unchanged by temperature, of importance mainly in instruments. Wood has very low expansion along the grain but high across it. The non-ferrous metals have coefficients approximately twice that of steel apart from titanium alloys, which expand less than steel.

Thermal and electrical conductivity is highest with pure copper (apart from gold and silver). They are generally high in pure metals and greatly reduced in alloys. Electrical resistance rises with temperature, roughly doubling in 200 °C, though constantan or Eureka are constant-resistance alloys very useful in electrical work. What is not always appreciated is that at sub-atmospheric temperatures the electrical resistance of metals falls strikingly; at liquid-nitrogen temperature it can be one-tenth of the room-temperature value. This effect is used in research work but not as yet in industry. Near absolute zero superconductivity sets in.

Maximum service temperature is governed by strength and chemical conditions. In the absence of oxygen and other reactive gases tungsten is outstanding; in some lamp filaments it is taken remarkably close to its melting point; hence high-performance bulbs are delicate when hot. Carbon and molybdenum can serve as heating elements if oxidation is prevented. Chromium has been found rather brittle, possibly because of dissolved nitrogen, etc., but deserves closer attention, perhaps combined with molybdenum. It has a high melting-point and forms a strong oxide skin. Alumina and other refractory compounds are well-known high-temperature materials, but some ceramics become electrically conducting at high temperature.

Where loads are severe tungsten carbide is used, and is a highly successful material for cutting-tools, wire-drawing dies, ball-bearings, valves, pivots, etc. Granulated and incorporated in a lower melting-point metal, it forms a weld-on wear-resisting deposit. For lower temperatures there are tungsten-bearing high-speed tool steels, maintaining good strength and hardness up to over 600 °C. Where hardness is less important and oxidation resistance is vital, a series of high-nickel gas-turbine materials are available for temperatures of 700 °C and over, depending on stress and permissible creep.

Most metals lose strength seriously around the midway mark between 0 °C and the melting point. An interesting exception is aluminium particle sheet, sintered and rolled from powder, which maintains its properties remarkably well up to 400 °C. The sintered structure may be responsible for this, noting that tungsten carbide components are also made from sintered material. Since sintering is coming into very general use for cheap parts, often usable directly, without machining, further advances may be expected in this field.

At the low-temperature end some interesting materials are available: gallium—tin eutectic melts at 20 °C, Wood's metal at 70 °C. This latter alloy is useful for filling pipes temporarily to prevent crinkling during bending, making temporary moulds or work-holders, mounts for groups of small punches or dies, and overheating protectors in electrical circuits. Rose's alloy melts at 100 °C and a range of alloys stretch from there upwards to lead—tin solder which sets at 183 °C, tin at 231.91 °C and lead at 327.3 °C, the last two being standard temperature-scale points. Richard Trevithick is thought to have been the inventor of the fusible plug for boilers that forms the fail-to-safety feature if the water level drops too low by melting and thus releasing the steam pressure before the fire overheats the metal. Printers use alloys of antimony which has the rare property of expanding on freezing, like water, giving casts which reproduce the type mould with great clarity and sharpness.

Lead- and zinc-based alloys are used for labour-saving forming dies for sheet metal press-work. They are cast to shape against a hand-beaten master, or as the matching other half for a die machined from steel. Their great advantage is the ease with which they can be melted down for re-use many times, possibly needing just an occasional dose of reviver alloy to make up for the loss of volatile or oxidisable elements.

Tantalum and zirconium are used in nuclear and gas-turbine service; cobalt alloys (Stellites) are used as tool tips and welded-on coatings.

Duplex or composite materials are possible problem-solvers. Amini *et al.*[36] review some current uses; older ideas are Sheffield plate, a silver—copper sandwich and scythes with a thin layer of hard tool-steel supported by wrought iron. Ransome's self-sharpening plough-share was chill-hardened on one face during casting by placing a metal heat-sink in the mould. This gives the cutting edge a hardness gradient; the soil erodes away the softer side thus keeping the edge sharp.

Now some of the more common materials met in engineering will be discussed. Properties are given at about 16 °C unless stated otherwise. Where a range of properties is given the first value is for the purest metal, the last for the most highly alloyed. Symbols and units are as follows (see appendix for more accurate conversion factors).

U.T.S.   Ultimate tensile strength, load/original cross-sectional area, $MN/m^2$ which is also $N/mm^2$

$E$   Young's modulus, tensile or compressive stress/strain, $MN/m^2$. Note that scientifically speaking $GN/m^2$ should have been used but can lead to errors. Also, conversion to other units can be the same for $E$ and U.T.S: to obtain tons/in.$^2$ divide by 15; to obtain thousands of lb/in.$^2$ multiply by 150; to get $kgf/cm^2$ multiply by 10. Note that kp means 1000 lb in the United States but in Germany it is the symbol for what is kgf in English usage

$\rho$   Density relative to water

M.P.   Melting point (solidus temperature in alloys)

$\alpha$   Coefficient of linear thermal expansion, parts per million per °C (p.p.m./°C)

$k$   Thermal conductivity W/m °C, the heat flow through a metre cube with 1 °C temperature drop, conductive heat flow being proportional to cross-sectional area and temperature gradient

$r$   Electrical resistivity, $\mu\Omega$ cm, the resistance of a 1 cm cube in microhms. This old-fashioned value is retained for its practical convenience

*Steels*
Steels are the most common group of materials in engineering, ranging from mild steels with 0—0.2 per cent carbon, through increasing amounts of alloying to

high-tensile, creep-resisting and stainless steels, with increasing cost due partly to the alloying elements and partly to the exactness of the additions and the need to exclude harmful constituents. Though mostly used in wrought form, most of the carbon and alloy steels are castable, giving lower strength values than the forged versions. All steels have $E \sim 200\ 000\ \text{MN/m}^2$, $\rho \sim 7.8$, $\alpha \sim 11$ p.p.m./°C, except for the stainless steels which have higher expansion, $\sim 16$ p.p.m./°C.

Mild steel has a minimum U.T.S. value of about $350\ \text{MN/m}^2$. Medium-carbon and low-alloy steels are used whenever higher strength is needed, especially in bolts and nuts, with U.T.S. in the range $500-700\ \text{MN/m}^2$. The stronger steels are not weldable without some precautions such as pre-heat, stress-relief, etc., as specified in makers' literature or metallurgy textbooks. High-alloy gear and spring-steels can be heat-treated to U.T.S. of $1500\ \text{MN/m}^2$ in some cases; many varieties can be surface-hardened, leaving a softer, tougher core. Even stronger steels are coming into use; fine wires with U.T.S. $2500\ \text{MN/m}^2$ have been known for some time.

Because the properties of steel are strongly influenced by small amounts of carbon (and also of nitrogen), the surface properties can be made very different from the main bulk. This has good effects as well as bad ones; the surface can be turned into a harder steel by exposure to a carbon-rich or nitrogen-rich environment in a furnace or salt-bath, usually followed by a heat-treatment cycle. The infusion of extra atoms induces a compressive stress, helpful in fatigue as mentioned in chapter 3. Against this, many heat-treatment processes allow some carbon to oxidise away. Decarburising is known to reduce fatigue strength; it seems likely that loss of atoms should leave the surface in a state of tensile strain. It does not need much imagination to see what stresses could be involved. Serious thought should be given to securing carburising rather than decarburising atmospheres in heat-treatment furnaces. This may give a side benefit in reduced scale formation.

Thermal conductivity of mild steel is about 65 W/m °C, down to 20 W/m °C for some alloys; the electrical resistivity ranges from 12 to $80-90\ \mu\Omega$ cm, and M.P. $\sim 1500$ °C.

Stainless steels require specific comment. There are basically three types: hardenable martensitic cutlery steels, the cheaper ferritic and the dearer austenitic types. The ferritic types contain little or no (expensive) nickel and tend to be less corrosion-resistant but are actually preferred in sulphurous atmospheres.

The dearer austenitic types contain 18–25 per cent chromium and 8–12 per cent nickel; they also contain stabilisers to protect them from carbide deposition during welding or brazing which would lead to rapid local corrosion (weld decay); 1 per cent of titanium or niobium or 2 per cent molybdenum are used as stabilisers, in conjunction with low carbon content. For full details, makers' literature or standard specifications must be consulted.

An extremely important aspect of design in steel is liability to low-temperature brittle fracture. Ferritic steels have a crystal system which has a transition temperature below which a crack propagates fast, in a brittle manner, while above this temperature the steel behaves in a ductile way. For many steel versions this transition temperature is between −5 °C and +5 °C. It escaped notice for many years since laboratory work is generally carried out above 15°C and there was no particular reason to suspect a drastic change in properties. The transition temperature can be lowered to a required level by alloying additions, though at some extra cost. Stabilised austenitic stainless steel has a different crystal system which seems to be free from this trouble, as are most non-ferrous metals.

The austenitic stainless steels have U.T.S. ~700 MN/m², with a low yield point when annealed and excellent ductility as a result of their progessive rate of work-hardening, which also makes them difficult to machine, especially if fine cuts are taken where rubbing plays more part than in deep cuts. Their high thermal expansion was mentioned earlier; their thermal conductivity is low, $k$~16 W/m °C, which is useful in kitchen tools. In cryogenic work they are almost regarded as heat insulators. Being non-magnetic, they cannot be gripped by magnetic chucks in the workshop. Electrical resistivity is about 70 $\mu\Omega$ cm. They are readily welded provided air is excluded, for example, by argon-shielded arc, seam or spot-welding; M.P. ~ 1400 °C.

Ultra-strong steels, U.T.S. over 1600 MN/m², are available commercially but are expensive and difficult to machine, their strength approaching that of the cutting tools.

*Cast iron*
The next most common material, cast iron again is a family name. It comes in strengths ranging from 150 to 600 MN/m², obtained by control over cooling rate and by alloying. It contains graphite, for reasons presumably known to the reader. The melt contains 2–4 per cent dissolved carbon, giving a low solidus temperature and good fluidity. If we chill the melt rapidly, we obtain a hard, brittle material which is used to provide local wear-resistance. It is, however, too brittle for most bulk purposes. If we cool the melt slowly, equilibrium is obtained with most of the carbon separating as graphite, normally in flake form, giving grey flake iron. The flakes weaken the metal but provide internal damping and also good lubricating properties when the graphite is exposed by machining. $E$ is very dependent on graphite content and is much lower than in steel; $\alpha \sim 9$ p.p.m. /°C. In fatigue cast iron is not strong but at least it is not notch-sensitive, being full of notches already which also make it brittle.

Brittleness can be removed by heating the castings for forty-eight hours at about 860 °C. The graphite collects in roughly spherical groups, giving malleable cast iron. Chilled (white) castings can also be made malleable in this way but take longer. With large castings it is more economical to ensure spheroidal graphite in the first place by appropriate alloying and cooling or by 'seeding' the melt with nuclei just before pouring it. Iron produced in this way is often called ductile cast-iron in order to distinguish it from malleable cast-iron; the terms seeded iron, nodular iron, s.g. (spheroidal graphite) iron, and Meehanite (a trade name) are also used.

All the following materials tend to resist atmospheric corrosion to some extent.

*Aluminium*
Aluminium and its alloys are of course much lighter than steel, $\rho \approx 2.8$. Unfortunately $E$ is also lower at about 70 000 MN/m². $\alpha$ is high, 24 p.p.m./°C for the pure metal, 19.5 for Lo-ex, a low-expansion alloy intended for engine pistons; $k$ is high, in the range 240 to 160 W/m °C. 'Pure' aluminium usually means 99.5

45

per cent pure; it is very corrosion-resistant, and is used for food-wrapping foil and also for electrical conductors where it is superior to copper on weight and price for the same resistance but is less easily joined. There is an opening here for aluminium wire coated with copper or brass, or even with the more expensive tin. Overhead lines often have an inner steel core for strength. $r$ for aluminium is 2.82 $\mu\Omega$ cm, only 5 per cent more than the 'super-pure' 99.98 per cent metal. It is remarkably soft and ductile, beginning to stretch at stresses of about 30 MN/m$^2$ and work-hardening gradually to U.T.S. 80 MN/m$^2$. M.P. $\approx$ 660 °C.

The numerous alloys can be said to fall into a general-purpose range of corrosion-resistant, weldable and/or castable alloys containing manganese, magnesium, silicon, etc., which have a strength range of 220 to 300 MN/m$^2$ without heat-treatment, merging into a stronger range of alloys with copper, nickel, etc., most of which are not weldable and are less corrosion-resistant than the general-purpose alloys but have U.T.S. values up to 600 MN/m$^2$. Suppliers' literature must be consulted for details but it can be said here that heat-treatment is quite different from the usual steel hardening and tempering. Softening is by solution treatment at temperatures approaching melting, so that good temperature control is needed, followed by quenching. A subsequent low-temperature heating produces hardening. Some rivet alloys are arranged to harden at room temperature and have to be kept chilled to keep them soft until use. M.P. $\sim$ 625 °C. The most usual alloy for castings is one with about 11 per cent silicon and various minor additives for grain fineness, etc.

## Brass
Brass is another widely used family; common machine brass at 70/30 or 65/35 copper-to-zinc ratio is easy to machine and easy to solder, and hence it is widely used in electrical and other instruments. $E \sim$ 100 MN/m$^2$, $\rho \sim$ 8.5, $\alpha \sim$ 19 p.p.m./°C, $k \sim$ 110 W/m °C, $r \sim$ 6.5 $\mu\Omega$ cm. In the annealed state brass is very soft and easily damaged but when it is work-hardened the U.T.S. can exceed 400 MN/m$^2$. In water and some refrigerants brass is liable to lose its zinc and become weak and porous. M.P. $\sim$ 900–950 °C.

## Bronze
Bronze was originally the term used for copper–tin alloy, perhaps produced from ore conveniently found ready-mixed; this of course dates back to the Bronze Age. Now we have aluminium–, beryllium–, manganese–, nickel– and phosphor-bronze, as well as naval brass and gun-metal. Most are highly corrosion-resistant; phosphor- and beryllium–bronze are used for spring contact-carriers in electrical work. Some aluminium– and beryllium–bronzes can be heat-treated like steel, the latter to U.T.S. 1400 MN/m$^2$. Since all have $E \sim$ 125 000 MN/m$^2$, this gives some of the highest elastic deflections obtained in metal. Some brasses and bronzes form good sliding pairs with steel. Although bronzes can be used to provide corrosion resistance, they are more expensive than stainless steel which is equally suitable in many cases.

## Copper
Copper is used in piping, electrical work and where its thermal conductivity is essential. It is highly ductile but not strong, particularly in fatigue. $\rho$ = 8.9, $E$ = 125 000 MN/m$^2$, $\alpha$ = 17.7 p.p.m./°C, $k$ = 400 W/m °C, and if the copper is 99.94 per cent pure and oxygen-free, as used in electrical work, then $r$ = 1.7 $\mu\Omega$ cm

(1.56 at 0 °C). M.P. = 1083 °C. U.T.S. for dead-soft copper is $\sim$ 200 MN/m$^2$ but yielding can commence at 40 MN/m$^2$ stress. When work-hardened copper is slightly less conductive but U.T.S. $\sim$ 300 MN/m$^2$. For some uses the oxide-containing tough pitch copper is preferred, but not in hot atmospheres which reduce the oxide and leave the metal weakened by porosity.

*Titanium alloys*
Titanium alloys are somewhat costly but find a use in high-performance engines, etc., where their exceptional strength-to-weight ratio is important, though $E$ is only about 120 000 MN/m$^2$. $\rho$ is 4.5, $\alpha$ is 7 to 9 p.p.m./°C, $k$ ranges from 120 to 50 W/m °C and $r$ from 80 to 170 $\mu\Omega$ cm. U.T.S. can be 1400 MN/m$^2$ with heat-treatment but these alloys are very notch-sensitive under fatigue loads.

*Magnesium alloys*
Magnesium alloys are the lightest alloys in common use; $\rho \sim$ 1.8. Wrought alloys can be heat-treated to U.T.S. up to 600 MN/m$^2$, so on strength-to-weight they beat even titanium alloys. The more usual cast alloys are of U.T.S. around 200 MN/m$^2$. $\alpha$ is high, $\sim$ 26 p.p.m./°C, $E$ is low, $\sim$ 50 000 MN/m$^2$, $k \sim$ 80 W/m °C, $r$ is in the range 4 to 15 $\mu\Omega$ cm and M.P. $\sim$ 650 °C. For many purposes aluminium does the same job at lower cost; magnesium alloys have the advantage where buckling is the main limitation; (their first large-scale use was in airships), and also where minimum thickness is governed by casting technique or by heat-flow.

*Zinc alloys*
Zinc alloys are used for their ease of diecasting to accurate (consistent) size, combined with satisfactory strength, stiffness and toughness. Unlike aluminium, they electroplate easily for decorative purposes. $\rho \sim$ 6.7, $\alpha$ is high at 28 p.p.m./°C, the remaining properties being very similar to brass apart from M.P. $\sim$ 400 °C. Only the proper alloys should be used owing to the high sensitivity to some impurities which in the distant past gave zinc alloys a bad reputation. They are also subject to brittleness at low temperatures; this property is made use of in a flash-removal process.

*Plastics*
Plastics include the remeltable thermoplastics and the thermosetting ones, the latter being slower in production but more stable at elevated temperatures. As new better polymers are frequently discovered only very general remarks and data will be given. Even the relatively hard plastics have low $E$, typically between 1000 and 6000 MN/m$^2$, with $\alpha$ from 60 to 160 p.p.m./°C. The lower $\alpha$ value is achieved by 'filling' with as much short glass-fibre as mouldability allows. Most plastics are satisfactory only below 120 °C and also suffer stress relaxation, or loosening, under continuous load. Long-term strength is generally not over 30 MN/m$^2$ and the creep effect should be allowed for in design. At the time of writing polycarbonate plastics show the best general-purpose properties.

A remarkable series is PTFE and PTCFE, which show very low sliding friction even against themselves, which is rare, resist almost all solvents and chemicals, and stand temperatures up to 250 °C.

When long, strong fibres are laid up in line with the load and bonded with resin,

47

we speak of reinforced plastics rather than merely filled plastics. The main fibres in commercial use are oriented carbon, giving C.F.R.P., and glass, giving G.R.P. Extremely high strengths are obtained by highly directional lay-up and high compression to give a high fibre–resin ratio.

C.F.R.P. comes in two varieties, the use of high modulus fibre giving products of $E$ = 200 000 MN/m$^2$, U.T.S. = 1000 MN/m$^2$, and the use of high-strength fibre resulting in $E$ = 120 000 MN/m$^2$, U.T.S. = 1300 MN/m$^2$; $\alpha$ is reported as –0.7 p.p.m./°C. Across the lay the properties are essentially those of the resin, usually epoxy type.

G.R.P. laid up directionally gives $E$ in the range 10 000 to 20 000 MN/m$^2$ and U.T.S. from 100 to 200 MN/m$^2$; the basic fibres are said to be stronger than carbon ones, but they are weakened by handling. Values of $E$ and U.T.S. for random G.R.P. may be about half the above or less depending on the fibre content which varies greatly. Density values of about 1.8 may be expected with high fibre content for both C.F.R.P. and G.R.P.[37]

## Concrete

Concrete properties also vary greatly, $\rho \sim 2.4$, while $E$ varies with age, composition and load, 20 000 to 40 000 MN/m$^2$ being representative. U.T.S. is below 2 MN/m$^2$ but compressive strength can be from 15 to 25 MN/m$^2$. $k$ is around 1.5 W/m °C. $\alpha$ is 12 p.p.m./°C, about the same as for steel.

## Glass

Glass is used for transparency or corrosion resistance; for instance it has been used in flue-gas to combustion-air heat exchangers in power plant. Though a wide range of glasses exist, it is sufficient here to discuss common window-glass, with $\alpha$ = 5 p.p.m./°C, $k \sim 0.5$ W/m °C and borosilicate type, with $\alpha$ = 2 p.p.m./°C, $k \sim 1$ W/m °C. The latter resists thermal shock, for two reasons: heat is conducted through faster and also causes less differential expansion. $E \sim 75\,000$ MN/m$^2$. Strength depends very greatly on defects; very strong glass is obtained by etching away the defect-containing surface layer, 'toughened' glass is produced by chilling the surface of hot glass with air-jets to induce a compressive stress. Design stress figures are given[38] as 15–25 MN/m$^2$ for plate glass; 40 MN/m$^2$ has been suggested for toughened glass but the importance of scratches and invisible defects must be remembered. $\rho$ = 2.5, softening temperature $\sim 600$ °C for common glass.

A sintered glass containing 96 per cent silica is suitable for metal casting moulds up to 1650 °C[39]. This could be a useful substance for welders' backing-bars (see chapter 7).

## Timber

Timber, of density 0.6 upwards, $E$ from 8000 to 15 000 MN/m$^2$ along the grain, has strengths of order 30–60 MN/m$^2$ for soft-woods, 60–100 MN/m$^2$ for hardwoods along the grain. Permissible working stresses in building work are given as 7 to 20 MN/m$^2$ along the grain. Shear strength is $\approx 7$ per cent of the tensile strength. This is so different from metal values that it often becomes the critical condition. Cross-grain properties are much lower, but too variable to quote. Interestingly $\alpha$ is given as 2–3 p.p.m./°C along the grain, 20 p.p.m./°C across. The cross-grain properties are greatly influenced by moisture content.

# 7

# Static Joints

This chapter discusses both permanent and disconnectable joints. The most common permanent joints are made by welding, sometimes with a hot flame, more usually with an electric arc. The workpiece usually forms one of the electrodes; the other is guided by the welder and is either of tungsten, which is not intended to melt, or of filler material which forms part of the joint. Other methods use local heating within the joint region, sometimes combined with high pressure, the heat source being electrical resistance, friction, electron beam, laser light, impact or ultrasonic vibration.

Welds fall into three groups (see figure 59). Fusing together two edges, which must be adequately supported elsewhere, gives either a seal weld which is of very limited strength or a plug weld which, as the name implies, acts as a peg or rivet. Where the members overlap we have a lap weld. The strongest type of weld, however, is the butt weld. A full-penetration, fault-free butt weld, suitably heat-treated and machined down level with the adjacent parts, is generally agreed to be as strong as the parent unwelded metal for design purposes. This seems logical in that the disturbance due to such a weld is relatively small and can be reckoned equal to that due to minor casual defects in the base material.

Most welds are produced by hand-held consumable electrodes, coated with a suitable flux, or by long-life electrodes and filler rod. In both cases the operator produces a small pool of molten metal to which further metal is transferred from the rod. The edges of the pool cool rapidly by conduction into adjacent metal, which enables the welder to control the process. The electrical supply is stabilised to maintain a steady current despite variations in the arc length, also overcoming the natural instability of an arc which has negative incremental resistance, the current tending either to maximum or to zero.

Hand-welding is highly versatile; a skilled welder, though preferring down-hand work, can weld vertically and even upside down. The most important restriction arises from thermal effects: a long run causes large temperature gradients so that the weld tends to crack on cooling; also distortion of the structure is aggravated. When the material is thicker than about 8 mm it is usual to specify several successive runs and in many cases pre-heating of the parts. For details, manuals on

49

welding and/or data issued by makers of welding equipment or the relevant design codes should be consulted. In most cases the joint surfaces must be prepared and set up to give a definite small gap for the initial root run, which is generally imperfect underneath and is eventually cut away and rewelded from the reverse side. Some weld preparations are shown in figure 60.

The requirements are access for the first run of weld and minimising the volume of weld metal needed. Thus with rising metal thickness increasingly elaborate weld preparations become justified. In heavy plate work they are carried out by edge-planer or flame-cutting. Another requirement is similar heat-loss paths into both parts, particularly with good heat-conductors such as aluminium or in thin work where one side could melt off prematurely. Meeting this requirement also tends to get the weld away from a stress-raising shape-change.

Some butt welds are produced by machine. For thick straight runs which can be set up vertically, a fixed or travelling mould is built to confine the molten metal; the heat is developed within a bed of molten slag covering the joint, giving rise to electro-slag welding. Submerged-arc welding is a halfway house, the weld being set up horizontally so that only the underside needs confining by a manual weld or by metal or refractory backing-bars. The top of the weld is covered by a bed of granulated slag to prevent rapid chilling and to keep out oxygen and nitrogen.

The foregoing methods are restrictive design-wise but economical for thick work which in hand-welding would need a large number of runs. They are appropriate for the main joints of large vessels and structures. As the build-up of a component (sometimes referred to as a weldment) proceeds, access for such welding machines becomes impractical and freehand welding is resorted to, or welding by one-sided machines which imitate the hand-welding process.

Oxidation is minimised by flux applied with the filler rod, or by shielding with gas, argon being used for stainless steel or aluminium and carbon dioxide for carbon steel. After each run any oxide or flux should be removed and any cracks cut out and rewelded before covering them up with the next run. It is usual to over-fill the joint with so-called reinforcement, which actually produces weakening in the form of a geometric stress concentration of 1.3 to 1.5; in high-grade work this is ground off flush with the adjacent metal.

A fillet weld is relatively easy to produce but has several weaknesses which make it unacceptable for highly stressed regions. Consistently good root-fusion is difficult to produce, the finished weld is difficult to inspect at the important inner corner and there is discontinuity of shape. Guidance on safe stresses is taken from two sources, thus drawing on experience in widely different fields of service. The first is the A.S.M.E. code for boilers and pressure vessels, section VIII; the second is BS 153: 1972, for steel bridges[10,40]

The A.S.M.E. code insists on butt welds for vessels to hold lethal substances, and also for large very high-pressure vessels, etc.; various fillet welds are permitted for other applications. As an example, consider an end-cover attached to a cylindrical vessel as shown in figure 61. The shear stress at X—X, taking the whole circumference and counting both welds, must be below U.T.S./15 whereas tensile stresses in the vessel generally can be U.T.S./4. This cautious level is called for despite the relatively favourable load conditions of infrequent cycling, no reversed loads and the beneficial effects of the proof test which each vessel receives.

50

The A.S.M.E. code not only shows a selection of permitted design forms but also some forbidden ones, as a warning, so if you read the code, read it carefully.

In bridges, stress cycles are more varied; in large bridges dead-weight predominates and only small additional stresses occur due to loads, wind, etc., whereas in small bridges stresses can actually reverse as the load goes across, especially in shear members. BS 153 lists various classes of detail; the weakest class, class G, applies where fillet welds occur together with a sudden and substantial change in cross-section. Some examples are shown in figure 62.

Figure 63 shows the permitted stress levels under various forms of load variation. Numerical values are not given because BS 153 covers several grades of steel; moreover the loading used in the stress calculation contains a safety margin; thus actual numbers may mislead. The important point is the ratio relative to steady loading and unwelded metal. The implied stress concentration factor is around 2 for $10^5$ cycles, 5 for $10^8$ cycles of intended safe life, as can be seen by comparing the solid and dashed curves.

The horizontal axis of figure 64 is a parameter which enjoys the not very specific name of 'stress ratio'. In the fatigue business this term is used to describe the extremes of a loading cycle in terms of the highest load. Thus a stress ratio of 1 is a steady load, a stress ratio of 0.5 means load cycling between half-maximum and maximum, a ratio of 0 means load cycling from zero to maximum in one direction; for complete reversals of stress the value becomes $-1$. (This aspect will be raised again when discussing figure 78.)

Joints, reinforcements and suchlike which are designed with a progressive well-graduated change, especially at the tip, are obviously better, being well established in brazing and gluing work. In welded work they are often used but have not been investigated sufficiently to establish safe design rules. Figure 65 illustrates this type of joint; the change of cross-section is obviously gradual, moreoever the fillet weld and heat-affected zone occupy only a relatively small proportion of any one cross-section. (Some of these forms of detail allow crevice corrosion to creep into the joint and may not be thought suitable for outdoor work.)

It is emphasised that the welded joint affects the main structure even if it is not stressed in itself, for example, where a lifting lug is welded on, because of the discontinuity of shape and the effect of welding on the base metal. The stress levels as given in the standard must not be exceeded in the main structure near the weld, nor in the weld itself. Therefore welded attachments are wherever possible confined to regions of low stress; reinforcements are carried well beyond the highly loaded region. Thielsch[41] is well worth reading on this subject. He shows enough failures to scare one off ever designing anything.

In general machinery design the problem is often ameliorated by the need to design for stiffness, stress levels being quite moderate. Difficulties are most likely to arise from failing to visualise the loading three-dimensionally; in this respect the use of simple cardboard models is helpful.

Stress estimates in welds are similar to other stressing problems; the smallest cross-sectional area for failure must be found, usually at the throat as defined in figure 59 (p. 52). Where fillet welds are in tension as in figure 65a, the equivalent structure suggests that the force in each weld is $F/\sqrt{2}$, acting in tension on the throat area. Ancient custom ignores the angle and calles it $\frac{1}{2}F$. This form of detail

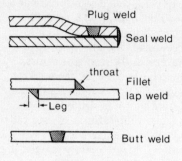

Plug weld

Seal weld

throat

Fillet lap weld

Leg

Butt weld

Figure 59

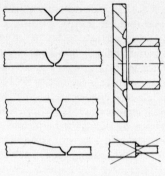

Figure 60

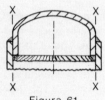

Figure 61

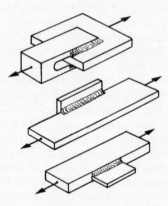

Figure 62.
Low-class welded details.

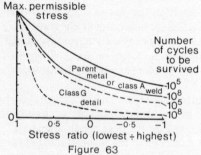

Max. permissible stress

Number of cycles to be survived

Parent metal or class A weld $-10^5$

$-10^8$

Class G detail $-10^5$

$-10^8$

0

1      0·5      0      −0·5      −1

Stress ratio (lowest ÷ highest)

Figure 63

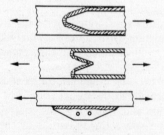

Figure 64

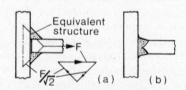

Equivalent structure

F

$\frac{F}{\sqrt{2}}$      (a)      (b)

Figure 65. Tee welds

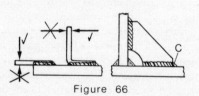

C

Figure 66

52

is not favoured where stresses are high, a full-penetration *T*-butt-weld as in figure 65b being preferred.

Important design points for fusion welds, in addition to general disposition and correct stressing, are access for the welder, especially to the underside, weld preparations, ease of dressing off excess weld (machining can be cheaper than manual grinding if the set-up is convenient), access for ultrasonic tests and X-ray photographs. Sometimes a later weld obstructs inspection of a previous joint.

The working drawing should specify the filler material which is often different from the parent metal, for example the well-known aluminium/3 per cent magnesium alloys need filler with extra magnesium to counter the loss by evaporation; in some cases the filler may contain additives to encourage fine grain-structure. Any pre-heating and particularly subsequent heat-treatment must be adequately specified. All these can with advantage be agreed with the fabricator in advance but must still appear on the working drawing, both for future repeats and in case the work is contracted out at the last moment to another shop. Letters are easily lost or ignored, and the drawing is the only document which is sure to go with the work.

In high-stress cases one should avoid a peel-stress. This is the name for a case where high tensile stress coincides with the end of a weld run. Two examples are shown in figure 66 together with the favourable and unfavourable directions of loading. If the load is in the unfavourable direction, we simply redesign the detail the other way round. A way of avoiding welding over and across a previous weld is also shown in figure 66. Note how the corner of the gusset is cut off at C to give a good fillet attachment. An acute corner would tend to melt and give an untidy fillet.

Resistance spot-welding, a very cheap and convenient mass-production method, is produced by clamping two parts between copper alloy electrodes and passing a heavy current, the contact force and current both being controlled throughout the cycle which can take less than one second. A lentil-shaped nugget of fused material is produced whose diameter is three to five times the thickness of the thinner part being joined. Design data are hard to come by; the static strength seems to be that of the surrounding metal, slightly annealed by welding heat. Available fatigue strength figures[42] do not state the actual nugget size but seem to imply a stress-concentration factor of around 3 in repeated one-way loading. It is thought that there might be some advantage in using annular electrodes relieved at the centre to give an enlarged periphery for a given weld volume.

To avoid a weak assembly it is helpful to be generous with the overlap and to stagger the spot welds, giving progressive load transfer and avoiding a line of weakness (figure 67). Multiple spot-welds are also a good precaution against an odd faulty spot.

Seam welds are formed similarly, using wheel electrodes and pulsing the current to give overlapping fused spots. They can be fluid-tight but in fatigue loading the possibility of a line of weakness should be considered. Another variant of the spot weld is the projection weld, where the spot size is governed by small projections formed on one of the components. Such joints are obviously weaker than proper spot welds and are used for attaching cable-clips, captive nuts, etc.

An intermediate case between arc welds and spot welds is the flash weld. It requires a machine set-up in which the parts are brought together then separated

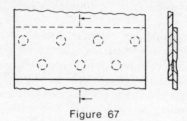

Figure 67

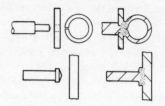

Figure 68. Flash and stud welds

Figure 69.

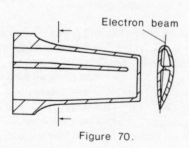

Electron beam

Figure 70.

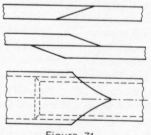

Figure 71.

54

again to form a short arc; when enough metal is fused the parts are squeezed together, expelling oxide, etc., into a bead or flash. To ensure a sound joint it is helpful to prepare one of the parts as a shallow point. This process is widely used, for example, to attach carbon steel shanks to high-speed-steel drill tips. It can even be used to assemble three parts at once (figure 68). The middle part is firmly clamped between the outer parts but not necessarily fused to them.

On small joints the process can be done by a one-sided machine known as a stud-welding gun. A cable is connected to the main workpiece to take the return-current while the gun flash-welds studs, cooling fins, captive nuts or small brackets on to the surface. The surface must either be supported or of sufficient mass to stay in place during the cycle which may take perhaps half a second. In small-size work, a satisfactory cycle can be produced by connecting both parts to a charged condenser and bringing them together rapidly.

Friction-welding is in effect similar to flash-welding, for parts of matching cross-section. The parts are moved over each other rapidly while pressed together with a substantial contact force until the whole surface is melted. On circular work, one component is rotated about its axis; on non-circular work it is orbited with a small circular motion at high speed. After a suitable time the relative motion is stopped rapidly but the contact pressure is maintained or even increased while the molten surface solidifies. This process seems to ensure that the whole surface is fused which makes it suitable for relatively large work, but it tends to be slower than electric flash-welding.

Electron-beam welding is a highly energy-intensive method, equalled in this respect by the laser and possibly condenser discharge, but it is the most amenable to automatic control. Best focusing needs a vacuum chamber; the focused beam produces a small region of vaporised metal. If the work remains stationary, a hole is produced. If the beam is pulsed rather than continuous, this hole can be very small. If the beam is moved relative to the work, the vaporised metal solidifies immediately the beam has passed; occasionally cavities are trapped in the work; there is a slight shrink-mark at the top and a bead of flash below. Usually no filler is used although it can be pre-placed in some cases, and there is no initial gap; therefore one expects a small shrinkage or a little residual stress but very much less than the local stresses in arc-welding. Dissimilar metals can be joined readily.

Electron-beam welding is not an exotic, expensive process; for example, it pays to use it for tipping tough carbon-steel hacksaw blades with high-speed-steel cutting edges. Groups of gearwheels can be cut first and then assembled by electron-beam welding, cheapening production and producing a shorter, stiffer gearbox design. The examples in figure 69 were provided by the Sciaky company of Vitry sur Seine, France.

The gearwheel A is built up from two different alloy steels at ten parts per minute. The teeth are finished before welding; only circular machining and sur-face hardening are needed after welding.

B shows a gearbox planet carrier combining mild, medium-carbon and alloy steels, produced at four parts per minute. Forgings and simple pressings replace a complex casting and save much machining.

The beryllium—copper diaphragm C, only 0.127 mm thick, is welded to a nickel disc D 2.3 mm thick to give a gas-tight capsule.

The nitrided mild steel pressing E is welded to the alloy steel camshaft F using circular deflection of the electron beam, at twenty-four assemblies per minute.

A particularly interesting facility of electron-beam welding design-wise is the ability to produce obstructed or interrupted (multiple penetration) seams, for making hollow bodies with stiffeners, cooling ducts, etc. (figure 70). Only straight beams have been used so far but it may be possible to curve the beam magnetically without undue loss of focus. The use of a vacuum container for the work is not absolutely essential; the beam can go through a short distance in air without excessive loss of energy or focus. This increases versatility on large work.

A laser beam can supply energy in a form as concentrated as the electron beam and should be able to produce similar work, without the need for a vacuum. However, it cannot be steered electrically like the electron beam; also maximum penetration is less.

Sources of welding heat are dangerous to the eyes. Even in spot-welding shops safety spectacles should be worn to ward off hot metal splashes and ultraviolet rays. Arc-welding is a very powerful source of these rays; welders must always use goggles but all rays are harmful and welding positions must be surrounded by screens whenever possible. Far more danger comes from laser beams; they are normally so small as to be unnoticed. The eye focuses laser light to a very small intense point and small spots of permanent damage can occur. Even an indirect beam reflected from glass, bright metal or liquid may be harmful. Not only must operators be protected by goggles absorbing the wavelength employed, but every effort should be made to prevent escape of the beam. People working with lasers have to undergo frequent eye-inspections to ensure that the protection is working adequately.

In brazing and soldering an intermediate metal is used to make a joint between two surfaces. Because a brazed or soldered joint is generally weaker than the components, a lap joint must be used to provide sufficient area, especially since the stress distribution is generally quite non-uniform. Only a scarf-joint or its equivalent produces uniform stress (figure 71).

The intermediate metal used in brazing and soldering is of lower melting point which renders the operation much simpler than welding, the whole joint or even the whole component being heated. The joining material is best pre-placed in the joint as a wire, foil or paint of powdered joining metal in a volatile carrier. To ensure a sound joint, the surfaces being joined must be freed from scale, grease, etc., to facilitate wetting. Fusion is either carried out in a furnace with a reducing atmosphere which is quite feasible in mass-production, or a flux is used to dissolve and displace any oxide, etc. The snag with flux is that most fluxes are hygroscopic and, once wet, corrosive. The only exception is resin flux which is used in soft soldering of particular types of surface, notably tinned surfaces and very clean brass or copper surfaces. The filler materials used in brazing and soldering are

(1) Copper, which forms a strong 'braze' between steel faces provided that the joint is thin, less than 0.05 mm being recommended. This material demands the use of a furnace with a reducing (hydrogen) atmosphere; the thinness of the joint also implies that the surfaces to be joined must be flat enough to fulfill this requirement. If the joint is thicker, the capillary action may be insufficient to fill the joint.

(2) Brazing alloys of melting point over 850 °C.

(3) Silver solders which cover a range of 625—700 °C.

(4) A range of alloys specially intended for brazing aluminium assemblies.

(5) Lead—silver, etc. alloys which melt at around 300 °C.

(6) Soft solders based on tin and lead, some of which deliberately have a long melting range during which they are pasty and wipable so that heavy joints can be built up.

Methods of heating include passing the work through a furnace, and localised heating by flame or by some of the electrical methods used in welding but at lower power and for longer periods to give time for the capillary flow to wet both surfaces. For soft soldering more gentle forms of heating are preferred, the soldering iron being the best known. Placing the work on a hot-plate at perhaps 300 °C for lead—tin solder is a good way of filling large joints which may be difficult to heat evenly. It is also convenient for large numbers of assemblies provided they are self-locating. Some electrical work is soldered by dipping the joint or by presenting it to a 'standing wave' produced by a pump.

Brazing and soldering can be regarded as demountable joining methods but since some diffusion takes place between filler and base it can require a higher temperature to undo a joint than was needed to make it.

The design of brazed and soldered joints presents few problems; the main points are that in furnace brazing relative slip between the parts must be prevented, the filler material must be in contact with the joint to ensure that it flows in as it melts and the gap must be suitable for capillary flow to take place.

If dissimilar parts are being joined, for example, brass to steel, it should be arranged so that the contraction in cooling down from brazing temperature makes the joint tighter, not slacker. For instance, joining a brass end to a steel tube should be as in figure 72a, not figure 72b.

Detailed advice is available in books on these subjects and in literature provided by the makers of soldering materials; much depends on local practice and equipment available. Particular care is needed in very fine-gauge joints since the solder affects the adjacent metal.

Riveted joints are another semi-demountable assembly method. Normal rivets are very cheap but it should be noted that 'blind' or pop rivets which can be secured from one side only by using a pull-through dolly to expand the rivet tail into a kind of head, can be as costly as screws.

The main advantages of a riveted joint are that it need not involve heat, it hardly ever comes undone and is correctly located. The disadvantages are weakening of the components by holes, the cost of providing the holes and assembly costs of the several parts.

In minor attachments the weaknesses to be considered are chiefly stretching of the component over the rivet head or shearing off the head itself; hollow rivets must not be too short (fig 73). For serious loads, rivets should always be in shear, preferably in symmetrical joints which use the rivet in double shear. A riveted joint is obviously weaker than the parent plate because of the reduced cross-section at the hole; if the loading is of fatigue character there is a further weakening due to stress concentration and danger of fretting; in lap joints there is a bending effect which can multiply the stress by four. Thus it is not surprising to find in

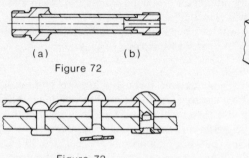

(a)                    (b)

Figure 72

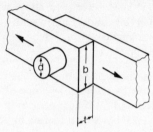

Figure 74

Figure 73

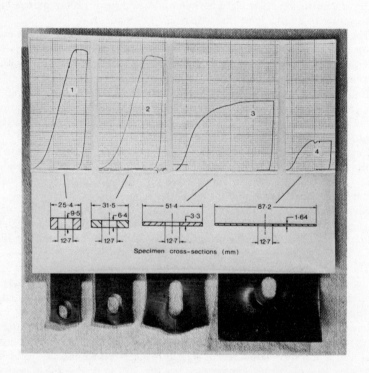

Figure 75

58

Heywood[14] that riveted joints in high-tensile aluminium alloys have a fatigue strength as low as 15 MN/m$^2$.

The obvious requirements are (1) enough cross-sectional area of rivets and (2) enough metal left at the weakest section of the plate. Let us consider a given load per rivet, thus fixing the rivet diameter, and see how big a plate goes with it. For simplicity let the rivet material and plate material be the same. The design stress values are taken from the A.S.M.E. code[10] mentioned earlier; they are U.T.S./4 for tension and U.T.S./5 for rivet shear. (The student will realise that since the nominal tensile stress is limited to U.T.S./4 the corresponding shear stress will be U.T.S./8. How do we justify the higher value of U.T.S./5? The answer lies in considering how well the rivet is supported; shear slip-planes cannot develop freely wherever they like as in a tensile test-piece situation.)

Using the notation of figure 74, plate strength and rivet strength are equal in a balanced design.

$$(b - d) \times t \times \text{U.T.S.}/4 = \tfrac{1}{4}\pi d^2 \times \text{U.T.S.}/5$$

Suppose we make the plate thickness $t$ equal to the rivet diameter; this gives

$$b = 1.628d \qquad \text{plate cross-section} = 1.628d^2$$

Instead, make $t = \tfrac{1}{2}d$

$$b = 2.256d \qquad \text{plate cross-section} = 1.128d^2$$

Next, try $t = \tfrac{1}{4}d$

$$b = 3.51d \qquad \text{plate cross-section} = 0.88d^2$$

The wider and thinner we make the plate, the more economical the design seems to get. This can be confirmed from common sense. The plate is weakened by the amount of metal drilled away; the thinner the plate the less metal is turned into chips, seeing that the rivet size is fixed by the load requirement. If we go too far with this process we must expect trouble at the contact point between plate and rivet, where the lines of force are crowded into a small cross-section. On a small, localised scale we could tolerate some yielding. To get a generally satisfactory joint, the code limits the mean bearing stress to 0.4U.T.S. This automatically avoids ridiculous extremes.

In our example, the load (which is $\tfrac{1}{4}\pi d^2 \times \text{U.T.S.}/5$) $\leqslant d \times t \times 0.4\text{U.T.S.}$, giving $t \geqslant 0.393d$. If a stronger rivet material is used or if the rivet is in double shear, the plate thickness should be increased appropriately.

What can happen when the plate is too thin is seen in figure 75. Four specimens of roughly similar material (mild steel) and equal minimum cross-section were tested in tension. The load—extension curves for each are shown. Note the long extension at roughly constant force as the circular hole extends to form an oval, giving high energy absorption. At peak load the tensile and bearing stresses were as follows.

| Specimen | 1 | 2 | 3 | 4 |
|---|---|---|---|---|
| Tensile stress (MN/m$^2$) | 541 | 510 | 362 | 131 |
| Bearing stress (MN/m$^2$) | 812 | 1020 | 1087 | 770 |

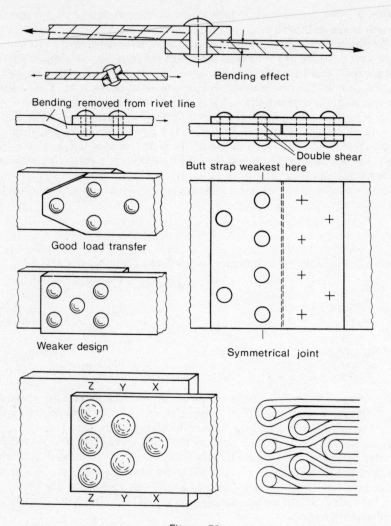

Bending effect

Bending removed from rivet line

Double shear

Butt strap weakest here

Good load transfer

Weaker design

Symmetrical joint

Z  Y  X

Z  Y  X

Figure 76

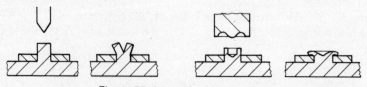

Figure 77. Integral rivets

A joint normally uses several rivets. If all the rivets are placed side by side the case is as for a single rivet, often aggravated by plate-bending effects since the load line tends to go through the shear-plane. (This can be reduced by a joggle.)

Efficiency is increased if the rivets are placed in two rows; then the plate is weakened by only one set of holes instead of two. If still greater efficiency is required, the rivets can be grouped in three or more rows. The size of rivets used is such that all the rivets together can take the working load without exceeding the permitted stress level. The plate thickness is worked out so that the full load can be taken at X–X (figure 76). At the next row, Y–Y, the plate is pierced by two holes but is no longer carrying the full load, having already transmitted one rivet's share to the other part. At Z–Z it is weakened still further but is relieved of three shares of load. This is another instance of progressive load transfer. The check for tensile stress needs to be made at each row though usually the first row is the critical one.

A traditional way of visualising this situation is to imagine the plate cut up into bands, one band looped round each rivet. If rivets of varying diameters are used, the width of each band would be proportional to the rivet's cross-sectional area in shear. The band idea also shows the importance of leaving enough metal at the sides and at the free edge of the plate. Cycle chains and suspension bridge chains show the same principle. It is usual to leave metal width of one hole diameter or more, especially at the rear. Design codes contain definite rules for edge distances, which vary according to circumstance. In timber, weak in shear, much greater distances are needed than in metal.

Rivets in shipbuilding are assembled hot, making it easier to form a smooth head and giving a tighter joint than cold riveting. Small rivets can be resistance-heated *in situ* and headed using the same method as spot-welding, the electrode being shaped to form a neat head.

In diecasting and plastic mouldings integral rivets can be formed on the parts, and these are spun or staked-over on assembly, hot or cold (figure 77).

Bolted joints come in two basic forms, shear and tensile. Where bolts are used in shear the principles of design are the same as for rivets, subject to the extra worry that a wrong bolt may be fitted so that the threaded part comes at the shear line, leaving only the root area available to resist the load. In structural engineering, friction-grip bolts are used which only act in shear as a last resort; they are in clearance holes and are tightened up to just about yield point thus clamping two members together so that the frictional force resists the load. Special design rules and construction rules apply to friction-grip bolts to ensure correct tightening.

In mechanical engineering, correct tightness is difficult to ensure. Tightening torque is a good guide only on new, clean, lubricated threads. Surfaces initially touch at high spots and eventually settle slightly, especially if thermal cycling occurs; on subsequent tightening, torque is no longer reliable unless full slackening and relubrication is permissible. In some classes of work, bolt extensions have to be measured; in studs or other cases where both ends are not accessible, a small hole is drilled down the centre of the bolt for a gauging device to note the elastic extension during tightening and also as a later check against settlement and hence loss of tightness. Some bolts are pre-loaded by heating or hydraulic jacking during tightening.

The difficulty with bolted joints is to determine suitable working stresses. In fatigue, bolts have been shown to be very weak[43]. Moreoever thread-rolled bolts were found to be much stronger than thread-cut ones but were remarkably weak if heat-treated after thread-rolling, perhaps owing to excessive decarburising. Figure 78 shows some typical fatigue strengths of bolt threads from an E.S.D.U. data sheet[43].·The values are plotted in terms of stress ratio which seems appropriate for design work where we usually know the maximum and minimum (or reversed) loads. The more fundamental Goodman diagram approach is best for research work since many testing machines have separate controls for the cyclic loads and the mean load level.

Fortunately bolted joints are usually well tightened and the flanges or components form a much stiffer load-path than the bolt. A load change produces a small deflection, giving a large readjustment of contact pressures and only a small change of bolt stress. If the stiffnesses can be estimated, a graphical procedure can be used to find stress changes and suitable initial loadings. This is described in several design manuals, for example in Phelan[24], (p. 154). See the appendix for a more detailed discussion. It has been noted[43] that the acceptable stress range is almost independent of the mean stress level.

A bolt can be made more resilient by waisting (figure 79). Note the centering ridges which prevent misalignment. A bolt lying at an angle to the hole axis is greatly weakened by bending effects. The centering ridges are kept clear of the head to avoid producing a severe stress-concentration there. Resilience can also be increased by lengthening the bolt and using spacer tubes. Belleville washers have been suggested for use as spacers, saving space and weight.

In practice some engine bolts are designed for tightening up to yield-point and the author has found big-end bolts which were visibly overstretched, presumably by some enthusiastic repairer, yet continued to hold despite the cyclic nature of the load. Such high stresses are not recommended for general service. For example, BS 2573 which deals with design of cranes permits maximum stresses for small bolts of 60 MN/m² for mild steel, with higher values for stronger steel and larger bolts. This implies that some corrosion and/or some over-tightening have been foreseen. If the reader looks up this reference he will find rather higher numbers but then the loads are modified by somewhat arbitrary service factors. The author has found that insurance companies prefer rather lower stress values.

The weakness of bolts comes only partly from the thread form; a large stress-concentration is due to the nut. On load, the bolt extends while the nut shortens; the pitch no longer matches exactly and the load concentrates strongly on the first turn of thread[14]. This can be almost completely relieved by a suitable undercut in the nut; owing to the loss of contact surface, a larger nut than normal may be needed. The undercut and tapered end ensure gradual load transfer, the nut and bolt both being in tension at the transition point, gradually changing to shear stress further up.

In steam-turbine casing joints where space is restricted, the problem is overcome by using nuts with a tapered thread. These are sometimes formed as cap-nuts with a reduced driving hexagon on the end[27]. The mode of action of these is fairly obvious from figure 80. The thread is cut to such a profile that under the design load, evenly transmitted, the shortening of the nut and the lengthening of the bolt

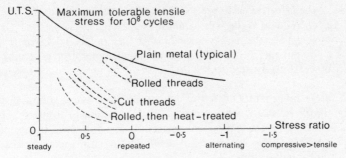

Figure 78. Fatigue data for bolt threads

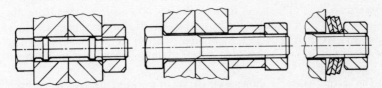

Figure 79. Increasing resilience

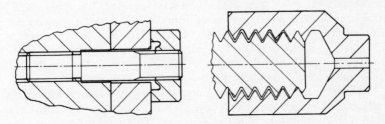

Figure 80. Reducing stress concentration

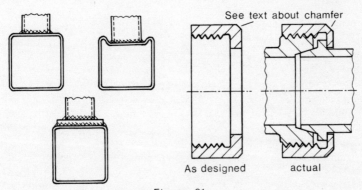

Figure 81

63

just cancel the effective clearance. Note that Poisson's ratio tells us how much the bolt diameter reduces as the bolt extends. The change in diameter of the nut is more complex, since the load is conical rather than purely axial. Where studs are set into a component, a counterbore about one diameter deep is recommended[44]. The action presumably approaches that of the relief groove in the nut mentioned above.

It is interesting that ISO (metric) bolts to BS 3692 have a good transition radius between head and shank which makes it most important to ensure that there is an adequate chamfer in the component adjacent to the bolt-head. In assemblies where the bolt could be either way round, it seems wise to ensure chamfered seatings for both possible positions.

The most common trouble in joint design is failure to consider load transmission or force flow. If we visualise load flowing like magnetic lines of force or like a fluid, then wherever we crowd the flow round sharp corners or through narrow passages we need to look out for high stresses. The analogy stops there; the stresses must be calculated or estimated by normal means. Two examples are shown in figure 81. The first, a detail for a tubular structure, is an obvious case of ignoring the beam condition which clearly calls for reinforcement, preferably tapered as in figure 64 (p. 52).

The second example shows a gas-fitting of the kind involved in the accident discussed in a Ministry of Housing and Local Government Report[45]. As designed it would have been quite satisfactory although a little on the slender side for a brass component unless high-tensile brass were specified. The weakness of the actual component is painfully obvious from the scale drawing, showing one of the virtues of proper engineering drawings. As a result of overstrain during fitting at some stage, a gas leak occurred and caused an explosion in a block of flats. The structural design of these flats was not explosion-proof, so the trouble was not localised as it should have been and produced a major collapse.

Note that in the Report there is a discrepancy over a corner chamfer. As drawn, it affects the strength yet on a photograph it is confined to the corners of the hexagon and is well away from the weak zone. This point is mentioned here to emphasise the importance of proper attention to detail. In this instance the conclusion is unaffected but this kind of error reduces the credibility rating of the experts and could be particularly damaging in court.

# 8

# Bearings and Seals

This chapter discusses joints with rotary or linear motion, mostly with sliding or rolling friction. It is, however, convenient to start with a short section on flexure bearings, useful where small movements are involved. In instrument work, crossed-leaf springs form a friction-free and slack-free pivot for angular movements of a few degrees. The arrangement requires at least three springs for symmetry. Any loads which tend to compress the springs must be kept small to avoid buckling.

Rubber bonded to metal gives rise to flexure bearings with somewhat imprecise location but with a wide choice of stiffness, partly by design and partly by choice of rubber composition. In compression the stiffness of a flat sandwich depends greatly on the ratio of bonded area to free side-area whereas the shear stiffness depends on total thickness. This is made use of in multi-layer bridge bearings which allow for horizontal movements between sections due to thermal expansion, etc., while deflecting very little in the vertical direction so that road level is maintained. Rubber bushes with voids left in the rubber have a similar effect. Rubber bearings of this type are preferred to metal rollers or slides which tend to corrode solid.

This controllable flexibility of rubber bearings is very useful in automobile engineering and has given rise to a proposal for taking care of the multiple freedoms and load directions at the base of helicopter rotor blades by a rubber mounting. Figure 82 shows a few basic flexure bearings.

One feature common to all flexure bearings is that they exert a restoring force when deflected. One exception to this is a relatively slim rubber mounting loaded or pre-compressed so much that it is on the verge of buckling like a strut so that the lateral restoring force is almost nil. This effect is more likely to be useful in experimental set-ups than in industrial applications[46].

Low-torque bearings for appreciable loads are needed in weighing machines; the most common mechanism used is the knife-edge, usually of 60 or 90° included angle, resting in a shallow V-shaped channel of about 150° angle (figure 83). In delicate laboratory balances 60° glass knife-edges are used, resting on flat surfaces, or else flexible suspensions. In heavier work, 90° knife edges and their supports are made of hardened carbon tool steel; if they are sharp (just free from burrs)

65

they can support 1000 lbf/in. length or 150 N/mm[47]. If greater loads are required the edges must be rounded slightly, thus increasing the resistance to motion and slightly reducing the positional consistency which ensures accurate weighing. Commercial weighing-machines using knife edges can be sensitive to 1 in 25 000 when new.

Watch and clock pivots, also used in many other instruments, use sliding friction between a steel shaft and a synthetic-jewel bearing. These materials are chemically incompatible so that there is no danger of local welding at high-spots. The friction coefficient may be appreciable but the *force* is concentrated on the smallest possible radius so that the *torque* is small. Because of this concentration, only small loads can be carried.

When greater loads are to be withstood, a ball-bearing pivot (figure 84) is believed to give very low friction torque, using rolling friction brought down to the smallest possible radius. Three balls are used to give proper kinematic location as explained in chapter 5. (As mentioned earlier, pure rolling is confined to cylinders and to cones with a common apex; cylinders are the limiting case, with the apex at infinity. In conical roller-bearings there is a resultant squeezing-out force which has to be resisted by sliding friction somewhere.) In the ball-bearing pivot, the position of the contact points on each ball is determined by force balance, so they are of necessity diametrically opposite and pure rolling is impossible; some degree of sliding must occur and lubrication is required.

While the bearing described above would be designed as required, most roller-bearings are bought complete with inner and outer races, the rolling elements and usually also a cage to keep them spaced evenly. Most versions are also available greased for life (more or less) and sealed with resilient barriers to keep the lubricant in and anything else out. Their load capacity depends mainly on the size of the balls or rollers; this determines the severity of the contact stresses. In radial loading, only two or three elements are effectively in the load-line, while axial (thrust) loading is shared out much more equably.

Load capacity has two aspects: a static bearing only used occasionally, such as on a swing-bridge, can fail by indentation, especially if the apparent load is augmented by a vibratory component which may well be of unpredictable size; on the other hand a bearing which moves frequently or continually will eventually fail as a result of fatigue. The probable life (number of revolutions) varies inversely as the cube of the load for ball bearings, or as the three-and-one-third power for roller bearings. The difference seems rather academic and could be due to the fact that ball-bearings rotate in all planes so that the work is spread over the whole surface of each bearing, while with rollers the end faces are not loaded. It is interesting that under slow continuous rotation the load can safely be made greater than the maximum static load, since indentation is prevented. Obviously this fact is only useful when we can be certain that the maximum load always occurs during motion and the machine is never stopped under full load.

The length of satisfactory life depends on the job concerned: a private car or a lift in a quiet hotel may run only 2 hours daily; some machines work 40 hours per week, others 168, quite apart from the speed range. Running speed in itself does not seem to affect the total number of revolutions before failure. The load capacity of bearings is given in makers' catalogues but the basis is in some cases $10^6$ revolutions, which amounts to only $5\frac{1}{2}$ hours at 3000 r.p.m., whereas in another catalogue the basis is more realistic, 1000 hours at 1000 r.p.m. This life does not mean that at the rated load and 1000 r.p.m. every bearing will seize up solid in exactly 1000 hours; the usual basis is a 90 per cent chance of still being in working order, though one manufacturer's catalogue uses a life basis such that *no* bearing will be excessively worn at the end of the calculated period if correctly fitted and lubricated. In a machine with many bearings it is important to overdesign considerably since the failure of any one bearing means a stoppage for repair, if nothing worse.

The cube law means that if we install a bearing whose capacity is 1.1 times what is needed, it should last $1\frac{1}{3}$ times the standard life; if we make it twice as strong as needed, its life will be eight-fold. In quiet situations a roller-bearing often gives an audible warning of trouble before it fails completely.

Figure 85 shows, to scale, a selection of bearings to fit a 40 mm shaft. Only the heftier versions are shown; there are lighter, slimmer bearings available for most shaft sizes. On the other hand some makers offer extremely strong taper roller-bearings of outside diameter almost three times one shaft diameter. (Note that in the diagrams the cages which keep the balls or rollers spaced out have been omitted for clarity.) The load capacities shown are calculated for survival, on the usual 90 per cent basis, to $10^9$ revolutions, which is one year's non-stop running at 2000 r.p.m. The radial loads and the thrust loads are to be understood as alternatives, acting separately. When, as is often the case, both types of load occur together, makers' data must be consulted. The general rule is that moderate thrust loads do not impair the permissible radial loading because of improved load-sharing; beyond a certain point this no longer helps much and the combined load becomes the ruling factor. This brief explanation should help the reader in understanding the makers' elaborate design instructions.

There are several problem areas in mounting bearings correctly.

(1) Thermal expansion of shaft or case could impose severe loads on rigidly fixed bearings. On long shafts it is usual to ensure that one bearing only is used for axial location, the other being free to slide or of non-locating design, for example, a roller-bearing.

(2) Flexibility or initial misalignment of machine frame or shaft can occur, particularly in farming, textile or similar large but light-weight machinery. If this is liable to misalign a bearing *unduly* (see makers' data for permissible angles), self-aligning bearings are used or orthodox bearings whose outer races are made spherical and set in corresponding housings, lubricated to ensure that the bearing is willing to line up while under load. For some heavy work, bearings with barrel-shaped rollers and a spherically-ground outer race are used.

(3) Creep between bearing and shaft or housing must be avoided, otherwise severe wear can occur there. This is not easy to explain on paper. Consider an undriven car wheel (figure 86), with exaggerated clearance between the hub and the outer race of the bearing. Points A and A' travel at the same *linear* speed; when A has reached position B, A' is some way behind, at B'. When A has gone full circle, A' has only reached point C. This type of creep can be avoided, either by a press-fit or in some cases by adhesives.

The inner bearing stays stationary relative to the load and need only be a sliding fit on the stub-axle.

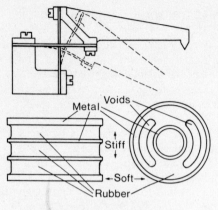

Metal — Voids
Stiff
Soft
Rubber

Figure 82. Flexure bearings

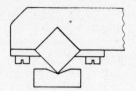

Figure 83

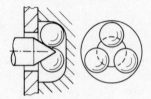

Figure 84

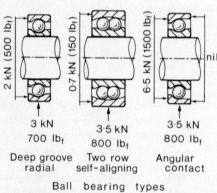

2 kN (500 lb$_f$)    0·7 kN (150 lb$_f$)    6·5 kN (1500 lb$_f$)    nil

3 kN            3·5 kN          3·5 kN
700 lb$_f$         800 lb$_f$         800 lb$_f$

Deep groove     Two row         Angular
radial          self-aligning   contact

Ball bearing types

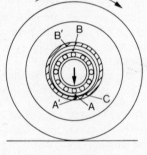

B' B

A — A — C

Figure 86. Creep

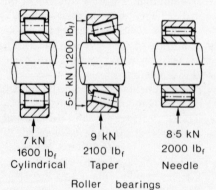

5·5 kN (1200 lb$_f$)

7 kN            9 kN            8·5 kN
1600 lb$_f$        2100 lb$_f$        2000 lb$_f$
Cylindrical     Taper           Needle

Roller bearings

Figure 85. Loads for $10^9$ revolutions

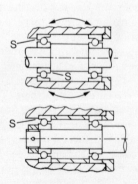

S

S

S

Figure 87

68

In the more usual industrial situation, the load-line is due to gravity, gear-tooth loads, belt-tensions, etc., but generally the shaft — the inner member — is the one which rotates, so the bearings are press-fits on the shaft and the outer race can be relatively free. In more complex cases both need fixing. For the correct amount of interference, makers' catalogues carry adequate instructions. Where both races are to be tightly fitted it may be necessary to order a special grade of bearing with enough initial slack to allow for this, since a press-fit causes quite a significant change of diameter. The housing should always be designed to leave a step for getting the bearing off again.

(4) A small problem is illustrated in figure 87. A wheel is often mounted between two angular-contact ball-bearings, adjustable for best clearance. If set up as in the upper view, rocking motion is possible, leading to wedging forces on the balls. To a lesser extent the same trouble could occur with taper-roller-bearings. The lower arrangement, which many readers will recognise as equal to the bicycle wheel lay-out, is free from this problem. If the hub can get hot, for example from braking, initial clearance is essential. The steps marked S are provided to aid in extraction as mentioned above.

(5) For very high speeds, special-grade bearings are obtainable made from parts selected for a particularly high degree of perfection, that is, roundness and concentricity of races, roundness and equality of the rolling elements. This grade is also used for machines where extremely true running is required, free from circular wobble or axial 'pumping' motion. The reason for using such bearings in high-speed work is the obvious one of avoiding large inertia forces. Another danger at high speeds is excess lubrication leading to power loss, overheating and, in extreme cases, cavitation damage. Fast-running bearings are usually pre-loaded against each other to prevent slack, with a spring-element to maintain pre-load despite wear or thermal expansion.

(6) In oscillating motion, for example, on an engine gudgeon-pin or in printing, packing or labelling machines, the inertia of the rolling elements tends to carry them on beyond the reversal point, especially in cases where at the instant of reversal the load is quite light. If this slippage is not prevented, the rollers or balls develop little flats and the motion gets slack and noisy. In fast-oscillating applications a permanent contact force is needed to give enough frictional grip so that rolling only, without slipping, is ensured. A pair of taper-roller or angular-contact ball-bearings can be used, pre-loaded against each other axially, or a specially-selected grade of bearing to secure this grip when installed.

Thrust bearings are similar in principle to roller-bearings but are laid out specifically for axial loads. Since they have zero strength in one direction, they are either used in pairs or kept together by a force applied through an angular contact bearing. Slack is not permissible; it would cause poor tracking and permit impact when the load comes on. The danger of overload due to thermal expansion is very severe in thrust bearings; a simple back-to-back arrangement which avoids the danger is shown in figure 88.

In large turntables the bearing size is often dictated by external conditions, the load per ball or roller being quite small. Bearings designed for such occasions are obtainable but the old-fashioned alternative of wheels running on a track or in a channel-shaped groove should not be overlooked, especially in plant where lubrication and sealing against wet or grit could become neglected in time, wheel

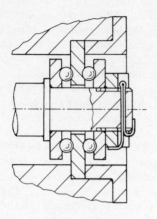

Figure 88

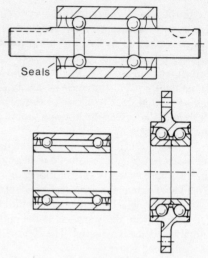

Seals

Figure 89

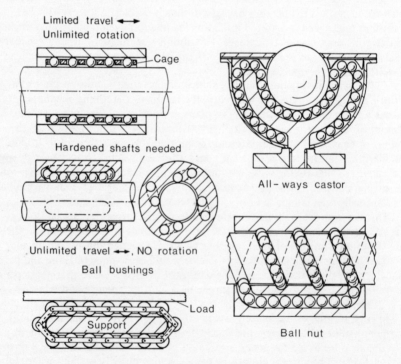

Limited travel ←→
Unlimited rotation

Cage

Hardened shafts needed

Unlimited travel ←→, NO rotation

Ball bushings

Load

Support

All-ways castor

Ball nut

Figure 90

70

bearings being easier to keep sealed than a large track.

Double bearings are a useful ready-made design feature. Three simple variants are shown in figure 89; they can be supplied sealed and greased for life, and also pre-loaded to be free from slack if required. Pre-assembled shafts and hubs with taper-roller-bearings are also available.

A few examples of non-circular roller-bearings are shown schematically in figure 90. Some versions use caged balls, some have free balls which push each other along, others use hollow rollers on a cycle-type chain, either as shown or with the rollers on the floor and the load connected to the support. A low-friction screw mechanism uses a recirculating ball-nut.

Sliding bearings are simpler, quieter and usually cheaper than roller-bearings but require several items for successful operation.

(1) A suitable supply of lubricant, for example, grease, oil, in some cases water, gas or certain non-inflammable synthetic fluids, or low-friction solids such as graphite, talc or PTFE (see chapter 6).

(2) Materials which make a good sliding pair, not prone to seizure.

(3) A sufficient heat-path or heat-removal arrangements to disperse frictional heat.

(4) If possible, a pressure-film to separate the surfaces. A successful pressure film, if thick enough to pass any grit through and clear all high-spots, could give an everlasting bearing if it were not for stopping and starting wear.

(1) Suitable oils and greases are available for most engineering situations. Their important properties are suitable viscosity both cold and at working temperature, surface-film strength, resistance to oxidation and in some cases water-repellence. Surface-film strength is to some extent inherent in oils but can be increased by additives. The action is generally that of a carpet of long-chain molecules, with one end fixed by polar attachment to the metal surface. These tend to have a temperature limit above which they de-sorb from the surface; where such temperatures are liable to occur locally, as in gearboxes, the oil may need oxidising additives to destroy hot-spots where peaks rub together.

An extreme case of surface action is molybdenum disulphide; this tends to form a film over the whole surface which is so strong that it can delay the running-in of new parts.

A curious difficulty of rapid wear is sometimes found in compressors for very dry gases and in vacuum apparatus; perhaps a little moisture provides a useful oxidising service. Note that graphite needs a little moisture to remain a good lubricant (hot carbon is a reducing agent).

Water is a suitable lubricant for some plastics on steel; water—oil emulsion is used in low-speed situations in some hydraulic plant but it should be remembered that oil protects against corrosion whereas water may encourage it.

Use of gas as the bearing fluid gives very low friction but to obtain self-energised wedge-films of gas needs high speeds and very fine clearances, making non-seizing surfaces particularly important since gas is not surface-active like oil. Air or gas bearings are at their best in externally-pressurised ('hydrostatic') form, ranging from large-area, low-pressure units better known as hovercraft down to small high-pressure bearings for instruments and dental drills.

Synthetic lubricants with special properties have to be used in some cases; non-inflammable fluids must be used in aircraft and in hydraulic systems in coal-mines, low-vapour-pressure oils are required in high-vacuum work, and radiation-resistant, high-temperature or oxidation-resistant fluids are used where appropriate. They tend to have worse surface properties and less resistance to cavitation than oils; on the other hand they generally have a lower viscosity index than oils, maintaining higher viscosity under local overheating (within limits) which is helpful in bearings. If, however, they are used in torque-converters or fluid couplings with turbine-like action one must check that viscosity is *low* enough at operating temperatures (at the blade surfaces which can be above bulk oil temperature). Some synthetic fluids have less molecular carpet action and greater liability to cavitation than oils. Hatton[48] contains many data on these fluids; Warring[49] is more recent but much briefer on this subject.

Points to watch with all lubricants are changes of properties in service, during rest or in storage. These can be due to freezing, tropical heat or humidity, radiation, or bacterial or fungus growth, particularly in tanks. Fungi can affect filters, contents gauges or level indicators. Overheating can occur during shut-down if an oil supply stops before the main machine is able to stop, or by heat-soakage from large hot masses. This may in some cases demand the incorporation of a heat-flow barrier.

(2) As mentioned earlier, chemically incompatible materials can make good sliding pairs. In most engineering situations a few other properties are required; successful bearings normally consist of a hard member, usually but not always the shaft, working against a soft material of duplex structure and much lower melting-point. In gearwheels, cams, etc., this is not feasible so both surfaces are hardened and well lubricated while rubbing speeds are kept low.

Duplex structure is favoured for several reasons: a soft base conforms to the shaft's shape and running position and lets accidental hard particles embed deeply so that they cannot become cutting-points to machine away the shaft. Also, the effect of fluid pressure and erosion by fine particles wears the soft phase preferentially, leaving the harder phase standing high; a well run-in plain bearing has a matt surface. This helps lubricant to stay in place during rest periods so that on start-up any metal-to-metal sliding is very short-term spatially and thus prevents the build-up of self-aggravating high-spots due to local thermal expansion. Such high-spot action is believed to be important in many friction situations, notably in brakes. The low melting-point limits the temperature of such hot-spots which keeps their height small and helps to avoid de-sorption of the molecular carpet-film.

(3) Heat is removed by two actions: direct conduction to the rest of the machine, eventually to a coolant or to atmosphere; this is aided by backing the bearing surface with a good heat conductor. The other, often the main means, is deliberate circulation of excess oil; sometimes it is possible to design the passages so as to encourage the fresh oil to sweep away hot oil which has passed through the bearing. Usually uncontrolled local mixing of fresh and hot oil is good enough.

Normally one pump supplies a number of bearings; then the danger arises that if one bearing wears slack it could take the majority of the supply and starve the others. To prevent this, the oil supply is restricted at each bearing by a metering jet; thus the slackest bearing gets some extra oil but the resistance of the jet ensures a basic line pressure to keep supplying the others. It is emphasised that this supply pressure is not the pressure which keeps the bearing surfaces apart; it is usually far too low to achieve this.

(4) A pressure film is either produced by pumping in oil at high pressure or, more usually, by letting the moving member drag oil into a converging wedge-zone. This of course means that on start-up there is a delay before the pressure builds up sufficiently to carry the load. In turbo-generators it has long been the practice to pump in 'jacking oil' at high pressure under the load-point to separate the surfaces before starting; pumping is continued until the wedge-film takes over (Church[27]). If jacking oil is not used, the action may be described as follows. Rotation starts by a small amount of rocking about the instantaneous centre at the rest-point A, figure 91, so that the contact point transfers to a well-oiled region next to A. This is sloping relative to the load-line so that sliding commences quite soon. Oil is dragged in towards the new contact point, producing a pressure acting upwards and also to the right, very soon forcing the shaft over towards the equilibrium running position. As the speed builds up, the pressure, though non-uniform as shown, becomes sufficient to oppose the load fully and the bearing then rides on an oil film considerably thicker than the molecular carpet. The clearance shown is grossly exaggerated, usual values being 0.001 times the diameter or thereabouts, depending on size and speed. (Locally the pressure in the divergent zone after the contact can become negative (tensile), soon dropping to about zero by cavitation of oil vapour and dissolved air.)

The basic design figure should logically be the least permissible film thickness, the choice of which depends on maximum possible misalignment and relative deflection, coupled with the degree of oil filtration. The pressure produced and hence the area needed to carry the load will then depend on the escape paths and the oil viscosity in the film. Here nature comes to our aid to some extent: at high pressures the viscosity of oil tends to increase; also the temperature is likely to vary through the film, so if oil escapes from the hottest, runniest layer the cooler layers form a reserve of cushioning power against short-time overloads.

If a bearing is narrow axially or has a circumferential groove dividing it into two narrow bearings, the outflow from these edges detracts greatly from the pressure-forming process. Even a (supposed) oil-supply hole in the load-bearing region becomes an oil-escape hole; indeed it was this discovery that started our knowledge of the wedge-film. If such a hole is needed, as for instance with the 'jacking' arrangements mentioned above, it is kept small and outward flow is prevented by suitable devices. There is, however, little advantage in making a bearing too wide; the benefits become offset by misalignment and elastic flexing which then become the restricting design condition. A little radial flexibility can be beneficial in relieving the highest pressures and reducing risk of rubbing at the point of closest approach and there have been experiments with flexible bands as bearings, but the danger of flexural stresses in the soft bearing material, which is liable to be weak in fatigue, sets a limit to this.

According to Church[27] the load per unit projected area (length x diameter) should not exceed 15 bar (225 lfb/in.[2]). The more recent *Tribology Handbook*[50] gives a detailed design procedure which for well-oiled, adequately cooled bearings of good aspect ratio (axial width greater than half a diameter) comes to the same conclusion (section A5 figure 5). If the bearing is narrower axially the side-leakage effect means less load-capacity per unit area.

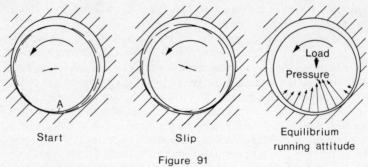

Start               Slip            Equilibrium
running attitude

Figure 91

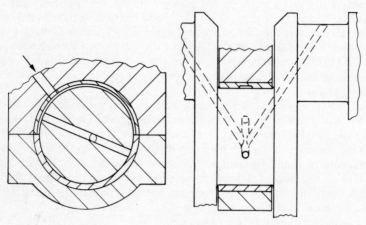

Figure 92

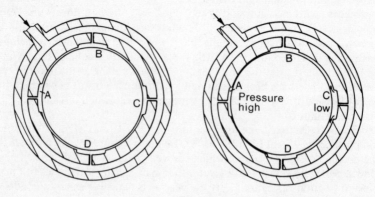

Figure 93

74

It may be noted that a greatly overdesigned bearing is to be avoided; first it wastes power, and second it may be unstable, the shaft centre precessing round and round at half the rotational speed (half-speed whirl), with roughness of running. The *Tribology Handbook* gives further details on this point also.

If the load alternates (as in many engines) the bearing can stand much greater loading since the gaps can refill with fluid at every cycle. At high frequencies there is insufficient time to squeeze the oil out again and the bearing operates under squeeze-film conditions. In high-speed four-stroke engine connecting-rods this action is more important than the rotational wedge film.

Readers familiar with internal combustion engines may have noted that the main (crankcase) bearings have a circumferential groove which supplies oil via the crankshaft to the connecting rods. This increases the leakage path as stated above and reduces load capacity. In most engines the highest loading is towards the bearing cap. This was taken advantage of in some aero-engines[51], confining the grooving to the less loaded part (figure 92). The crankshaft had two oilports so that continuity of supply was maintained while the highly loaded plain part had a wide uninterrupted surface. The same principle is believed to be still in use on one range of high-quality diesel engines.

Thrust bearings for major duties have a series of self-adjusting tilting pads or at least fixed sloping pads to provide a good wedge-film; even flat thrust-bearings have a small amount of wedge action, provided mainly by thermal distortion due to the normal temperature gradient from the hot film to the cooler machine parts.

Returning to the principle of using jacking oil to separate the surfaces – if this action is maintained continuously we have the hydrostatic or externally pressurised bearing, which can be of cylindrical, thrust-bearing or linear slide layout. The considerable pumping power consumption is offset by the advantages of just about zero wear and absence of starting-friction. In general it is not practical to use such a bearing as a single support point, for reasons of stability of position and angle. A typical layout consists of three or four support points, often recessed to give a pocket of appreciable area, each supplied through a metering orifice. Figure 93 shows such a bearing in the central, no-load position.

If no metering orifices were provided, the supply pressure would act equally on all sides whether the bearing were central or not, so the slightest load would push it right to one side, in other words the load capacity would be zero apart from a little wedge-film action at the lands around the pockets. With orifices, the action is as follows. Suppose an external load moves the shaft towards pocket A. The escape from C increases and the pressure at C falls; at A the escape is hindered and pressure in the pocket rises. We now see that a restoring force has appeared, giving the bearing a load-capacity and also a stiffness (force per unit displacement) which can be estimated with some confidence. It is usual to supply double the pressure needed to carry the load, dissipating the excess in the orifices. Detailed design procedures are given in Neale[50] and a P.E.R.A. report[52].

The simplest, cheapest lubricated bearing is the porous plain bearing, soaked in lubricant and often in contact with a felt pad, also soaked in lubricant, as reservoir. This form is sufficient for many applications. The lubricant is circulated through the porous surface by the action of the shaft dragging oil from the clearance side to the loaded side as in solid plain bearings, giving excellent local cooling. At high enough speeds and light loads the oil cannot escape fast enough through the pores and a true wedge-film can form. The chief limitation arises from absence of an external oil supply so that over-all heat dissipation (as opposed to the local one) is limited, being due to conduction only. This form is also discussed in Neale[50].

A few comparisons can be made about plain and rolling-bearings. Plain bearings are generally cheaper, quieter and more durable than rolling ones, especially at high speeds and with infrequent starts. Against this, they may need a powered oil supply and they will generally show higher friction, with coefficients up to 0.005. Roller-bearings inside machinery can have coefficients below 0.002, but self-contained, greased and sealed bearings may show appreciably higher friction than this, especially at light loads, since grease drag and seal friction are always there. Needle bearings, although compact and of high load capacity size for size, tend to have higher friction than large-roller bearings.

The remaining part of this chapter discusses rotary seals, followed by a brief note on piston-and-cylinder-type seals. Broadly rotary seals fall into two classes, the low-pressure type which keeps lubricants in and the rest out but with only slight pressure differences to cope with, and high-pressure seals as used mainly in pumps where the drive shaft comes through. Both are usually bought as proprietary products.

A typical low-pressure seal is shown in figure 94. A metal former, often fully covered with rubber as shown, is pressed into a machined recess in the machine. Moulded to it is a lip of a suitable rubber composition, lightly pressed against the shaft by a garter spring. The resilience of the rubber alone is not relied on for this purpose since rubbers are not too happy under tensile stress, tending to crack. Also rubbers, like plastics, tend to go slack (stress relaxation) and in many fluids they swell a little and lose shape. Rubber when dry is well-known for its high friction, hence its use in tyres, soles, erasers, etc. Now if a rubber seal is to stop the leakage of fluid while relative sliding is taking place, it has a self-contradicting task. Its success in remaining lubricated yet holding in the fluid depends on a surface-tension effect. When the motion stops, lubricant is eventually squeezed out from the rubbing surface and some wear is liable to occur at the next start. Some seals are skewed or have feed-slots or pockets to re-introduce lubricant as soon as possible after restart; others have the surface micro-roughened by a chemical 'weathering' process.

The author has shown that it is possible to make a rubber self-lubricating by introducing a substantial proportion of graphite[53]. A rubber with 25–30 per cent graphite, for example, giving a structure as shown in figure 95, proved to be remarkably wear-free. Lower graphite contents were hardly beneficial at all; excessive admixture weakened the rubber so much that although self-lubricated it crumbled from lack of cohesion.

At high pressures and speeds a rubber-type seal would not last long; mechanical face seals are generally used. The static face may be carried on a removable flange and may be seated in rubber to make it self-aligning for deflections of shaft or casing, temperature permitting. A clamp secured to the shaft has a driving key slotted into the seal member. This has a semi-static O-ring seal against the shaft and a spring to ensure initial contact. The fluid pressure provides additional contact force.

This force tends to be needlessly large. A partly balanced seal reduces the force, and the seal member slides not on the shaft but on the driving sleeve which

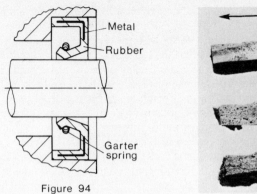

Metal

Rubber

Garter
spring

Figure 94

← 25 mm →

Figure 95

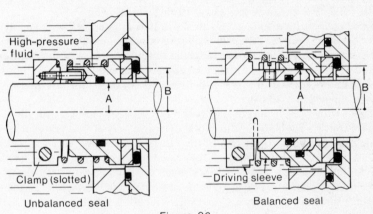

High-pressure
fluid

A

B

Clamp (slotted)

Unbalanced seal

A

B

Driving sleeve

Balanced seal

Figure 96

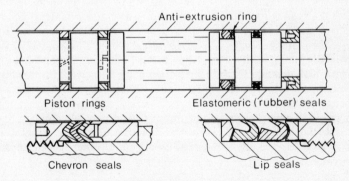

Anti-extrusion ring

Piston rings

Elastomeric (rubber) seals

Chevron seals

Lip seals

Figure 97

77

therefore needs an extra seal. The fluid pressure now acts only on the annular area between the seal radius A and the effective face radius B (which is approximately the geometric mean radius).[54] Figure 96 shows both types of mechanical seal.

The very nature of the seal means that the rubbing surfaces are under-lubricated. Some seals have various passages to promote partial lubrication and fluid circulation to remove frictional heat; nevertheless the surfaces need to be made of particularly wear-resistant materials, such as carbon or graphite against stainless steel or tungsten carbide. If water is being sealed, a non-conductor (ceramic or plastic) is sometimes preferred as one of the faces to avoid galvanic corrosion.

It must be recognised that any moving seal may leak a little, right away or later. If such leakage is unacceptable, the seal is backed up by another seal and the interspace is pressure-fed with a fluid which can be permitted to leak out and into the system. Sometimes lubricating oil can be used, sometimes an incombustible gas; in suction pumps the process fluid itself is used in this way to prevent the ingress of air. The same argument applies to vacuum systems, the interspace between seals being connected to rough vacuum. A flow indicator can be placed in the line to the interspace between seals to give early warning of seal failure.

Some reciprocation sliding seals are illustrated in figure 97, with greatly exaggerated clearance. They are generally self-energising; the machined piston-rings shown have gaps and are pressed outwards against the cylinder wall by their own springiness plus fluid pressure; axially they seal against one end face by fluid pressure alone. Care must be taken that inertia forces do not overcome this pressure; if they do the sealing is impaired.

Resilient material avoids the need for a gap. Rubber O-rings, with fibrous back-up washers for high-pressure work, or proprietary grooved rings are precompressed by about 10 per cent of their thickness. Axially, O-rings should be slack; a little rolling action helps to lubricate the sliding surfaces. For details see manufacturers' data; at low pressures a much smaller amount of nip may give better lubrication. O-rings of PTFE are compressed much less and therefore need close-tolerance surroundings; they are used where rubber is unsuitable, particularly at ultra-low temperatures.

Lip-seals are self-energising one way only and therefore have one-way pumping action; in a two-way seal which resists pressure in either direction they must be used back-to-back, lips pointing outwards, otherwise a high pressure could build up between them. Piston rings if tapered also show one-way pumping action.

These seals can work either way out, sliding against a cylinder bore or against an external surface. Unless their wiping action is very severe they tend to re-lubricate themselves at every stroke; chevron and lip seals are usually made of a self-lubricating composition comprising cotton, asbestos, graphite and waxes or greases for high-pressure work. Oiled leather is traditional for low pressures and modest temperatures. For ease of moulding leather seals are often of simple cup or hat-shape (without crown).

Where piston-rods are exposed to abrasive grit it may be necessary to cover them with expanding gaiters or at least to provide felt or other wiper-rings so that grit should not stick to the oily surface and be drawn into the mechanism, spoiling the main pressure seal.

# 9

# Damping, Mountings and Vibration

Vibration gives rise to two main types of design problems: one is that of isolating a body from random external inputs, the other, mounting a vibrating system so that as little vibration as possible is transmitted. The first is the transport (passenger or delicate cargo) problem; the second is the general machine or engine-mounting problem. A subsidiary problem is predicting and stopping odd component resonances. For the first case, particularly considering human cargo, there are some agreed standards of what inputs a person can accept for various lengths of time without undue fatigue. Unfortunately there are two sets of data in existence, sometimes confused with each other, one referring to drivers and passengers in vehicles, the other to people in buildings who have to stand, sit on relatively firm chairs or work on rigid tables. An internationally agreed set of standards for tolerable accelerations is given in Neale[50], section D3, but specific body resonances are not included.

For every sinusoidal frequency and length of exposure time there is a threshold of discomfort, a threshold of fatigue in terms of exposure time and onset of injury to health, roughly in the proportions 1, 3 and 6. A rough idea of vertical vibration effects is shown in figure 98. Published information refers mainly to vertical motion; fore-and-aft and lateral motions are less well documented but the tolerable levels are not very different from the vertical ones. An additional point noticed in aviation research and confirmed in automotive practice is that accelerations up to $1g$ *upwards* are much more unpleasant than the same accelerations *downwards,* presumably because upward ones magnify the normal gravity forces on the spine and on the ligaments by which the stomach and other organs are suspended, whereas downward accelerations reduce these forces. This is one reason why vehicle suspension dampers are biased to do their job of energy absorption predominantly on the down-stroke of the wheel.

In buildings, flexibility of the structure may amplify or attenuate the input vibrations, which can be both vertical and horizontal. The softness of vertically

79

flexible mountings is limited by tilting forces, especially in tall buildings; lateral flexibility is less restricted and is said to be very beneficial. A building mounted on springs rather than solid ground is somewhat like a ship or a car and needs a strong hull or chassis since it does not borrow rigidity from the ground. Care must be taken to fix gas, water and sewage pipes with enough freedom to accept the deflections without leaking or breaking. Particular attention must be given to the angle-fixed line-free buckling mode (figure 13).

For vehicles, the ideal would be to locate the passenger firmly in his visible surroundings and put these on a zero-rate, zero-inertia suspension. The aim with seats should be to provide uniform-pressure support but without over-all softness, for two reasons. Very soft seats give a feeling of uncertainty of position; also some soft seats are said to have resonant frequencies at around 3.5—4 Hz, which is one of the body-resonance regions[55]. Uniform-pressure seats may be filled with air, water or free-flowing granules. The extent of the support region is important; if it is not evenly distributed about the centre of gravity the occupant has to exert effort to keep in position. Pressure just under the knee is said to irritate; therefore thigh support should not be carried too far forward.

The springing and damping of a road vehicle is a compromise; the damping not only has to prevent repeated body oscillations but also has to keep the wheels on the road. Without dampers, the wheels are off the ground for a large part of the time (wheel hop); the damping force required to keep them down is sufficient to cause occasional discomfort. However, with low friction and non-linear dampers combined with soft springing (which may imply automatic level control), very good levels of comfort are now available.

The machine problem comes with three classes of input: single-shot as in drop-hammers, fixed-speed harmonic input but with starting and stopping to be catered for, and variable frequency. The first case has no resonance problem; what is needed is the softest possible suspension. Usually a very large foundation block is provided, resting on massive, expensive springs or a thin ribbed rubber mat. If larger displacements of the machine can be accommodated, it should be worthwhile considering a much smaller foundation, on much softer springs.

The ideal would be a constant-force device, with a servo-system to return the machine to its original level after each stroke. Figure 99 shows schematically a selection of devices which give almost constant force. Flat springs set up as Euler struts are shown alongside their practical embodiment, the Lord dockside fender[56], the struts being of rubber. The simple counterweight is sometimes appropriate. An academic curiosity is the flat spring flexed into an S shape as shown; its mid-point moves up and down with almost zero force. A gas suspension unit using a roll-sock diaphragm can be designed to give constant force if the supporting walls are contoured to vary the effective diameter $D$ so that $\frac{1}{4}\pi D^2$ varies inversely as the gas pressure; if designed isothermally, that is, for slow changes, it will not give constant forces for fast changes which affect the gas temperature. A vacuum capsule would give a force varying only with barometric pressure. Devices in which changes of linkage angle compensate for spring-force variations are used as pipe-supports to allow freedom for thermal expansion without loss of support.

The constant-speed machine is ideally damped by resonant or harmonic damping consisting of masses spring-suspended from the main body with a natural frequency equal to the operating one. Unless the machine is secured absolutely rigidly, its

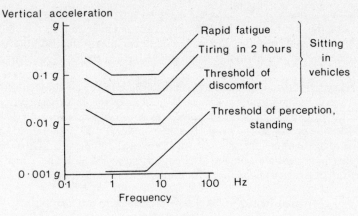

Figure 98

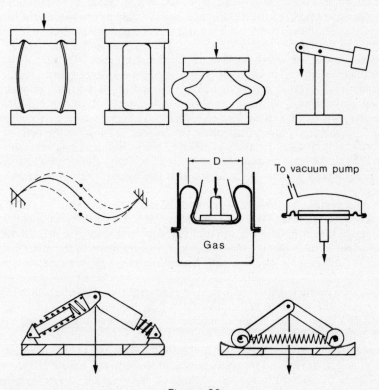

Figure 99

81

oscillations soon cause the harmonic damping system to oscillate in anti-phase, thus cancelling the input force. The difficulty if any lies in getting up to and down from running speed; this may demand an orthodox damping system additional to the harmonic one, since a harmonic damper takes several cycles to get into anti-phase as required.

Theory tells us how much of a given disturbing force gets through the suspension to the surroundings at various speeds. In most cases the disturbing force is of course not constant but also depends on speed which, as will be shown later, actually simplifies the problem. The property shown in figure 100 is the transmissibility, a rather odd term for something which can exceed 100 per cent. Other terms which confuse the student are gain and attenuation, expressed in decibels, being the logarithm of the transmissibility. The simplest relation to keep in mind is

$$\text{unit transmissibility} = \text{zero gain} = \text{zero attenuation}$$

This is entirely logical and forms a safe starting point, since $\log_{10} 1 = 0$. Ten decibels is a 10 to 1 ratio, 3.01 decibels is a 2 to 1 ratio, 1 decibel is a ratio of 1.26 to 1. In practice one must watch out because these ratios are sometimes used for force or deflection and sometimes for sound pressure or energy. Further confusion ensues since sound pressure is measured in decibels above $0.00002 \text{ N/m}^2$.

It can be seen from the graph in figure 100 that at speeds up to a certain value the suspension magnifies the disturbance, and it would pay to mount the machine as rigidly as possible so as to 'borrow extra mass', especially at or close to resonance. To obtain a satisfactory system the springing is designed so that the natural frequency is well below any possible sustained running speed. On start-up and shut-down the resonant region has to be traversed. Since resonance takes several cycles to build up fully, it may often be possible to pass through it quickly enough not to get high deflections. Unless one can be sure of a fast enough run-up and run-down, either damping or limit stops must be provided. The connections to a machine such as pipes, cables (and output shaft) must be designed to allow for the deflections.

Note that if classical damping is employed the modest damping value shown, 0.2 x critical, does not restrain the system at all severely, allowing up to three times the input force to get out. Yet at high speeds this damper transmits a great deal of force.

In most textbook cases and in many actual machines the major source of vibration is unbalanced mass. (If a machine of mass $M$ is disturbed by an unbalanced mass $m$, the rate of build-up of vibrations is proportional to $\sqrt{(m/M)}$. If both are tuned to the same frequency, the energy transfers from one to the other in $\sqrt{(M/m)}$ cycles.) If imbalance is used as a basis for the analysis, a far more instructive picture results.

Imagine a machine with an unbalanced mass of 0.1 kg, 100 mm radius (or amplitude). This would be typical of a fairly light single-cylinder engine or compressor or a larger, partly balanced multi-cylinder machine. It is mounted with a static deflection of about 22.5 mm, to give a resonant frequency of 200 r.p.m. If the imbalance is horizontal the case is exactly the same except that the static deflection is irrelevant and we speak merely of the stiffness. Figure 101 shows the actual force transmitted to the surroundings by this machine, at various speeds, with and without damping. At high speeds the damped suspension transmits quite large forces, mainly through the damper, but the undamped one settles down to a small, constant value. The physical reason for this is that a machine running well above resonant speed tries to keep its centre of gravity stationary. The imbalance means that the centre of gravity is not quite static in the machine but has a small amplitude or orbit. Suppose our machine had a mass of 100 kg, then

total mass × c.g. amplitude = unbalanced mass × its amplitude

making the centre of gravity amplitude 0.1 mm. The force getting out is essentially this displacement multiplied by the spring stiffness. (The academically minded could create a parameter, transmissibility multiplied by (disturbing frequency/natural frequency)$^2$ and call it somebody's number to add to the formidable list already in existence.)

Logically we should only apply damping when the oscillations are larger than the amplitude of the centre of gravity. We could fit a dash-pot with deliberate slack; alternatively we could make the dash-pot sensitive to amplitude and resonance, as proposed by Kindl[57] for motorcars (ignoring the need for wheel damping) using a spring-mounted seismic mass. At small amplitudes the fluid uses the large bypass passage in the piston rod. Whenever the mass has a large displacement the bypass is cut off and the dash-pot is fully active (figure 102). This idea, turned upside down both physically and functionally and with additional refinements, is used on a Japanese motorcycle[58].

There are some further complications in damping, arising from the six degrees of freedom to be considered. The relative simplicity of car engine mountings is made possible by the great variety of directional properties that can be designed quite readily into rubber pad mountings. Some systems combine soft mountings with check-straps or limit-stops against resonance and collision forces; using rubbers with sufficient hysteresis can avoid the need for separate dampers. Lindley[59] is recommended as a concise manual on the design of rubber components.

Resonance of components is discussed in various books on vibration and noise; here only a few points are selected. In whirling shafts, one point sometimes overlooked is the elastic deflection of the bearings, which is additional to shaft flexure and may be significant in very stiff, short shafts.

An oft-neglected resonance is surging of helical springs. A helical spring is an elastic medium with distributed mass and thus should have a calculable standing-wave frequency. For steel springs, it is $360\ 000d/(ND^2)$ Hz where $d$ is the wire diameter, $D$ the mean coil diameter (both in millimetres) and $N$ the number of coils. If $d$ and $D$ are given in inches, the frequency comes to $14\ 000d/(ND^2)$ Hz[60].

Occasionally the bell-ringing frequency occurs in thin hollow shafts and in pipes subject to high-frequency aerodynamic excitation. The fundamental mode has two lobes so that the cylinder goes slightly elliptical; higher modes have more. The fundamental frequency is $0.49ct/D^2$ Hz where $c$ is the 'bar velocity' or $\sqrt{(E/\rho)}$, more correctly in this case $\sqrt{[E/\rho(1-\nu^2)]}$; about 5200 m/s in steel and aluminium, 3600 m/s in brass; $t$ is the wall thickness and $D$ the mean diameter.

Noise can be structure-borne or airborne. Barriers against structure-borne noise are in principle the same as resilient mountings discussed earlier, and require proper design to ensure low transmissibility. It is of relatively little help in preventing noise transmission just to put a thin sheet of rubber under a noisy machine.

Against airborne noise there is only one perfect barrier, namely a vacuum.

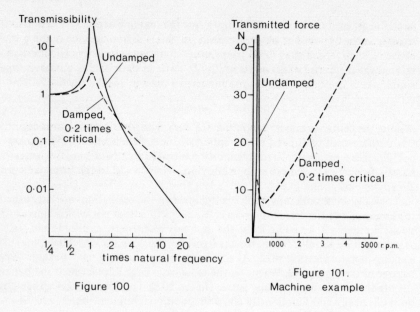

Figure 100

Figure 101.
Machine example

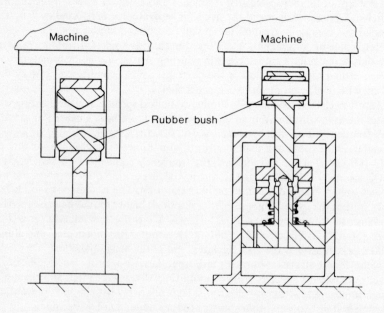

Figure 102. Machine damping ideas

84

Vacuum has not been used in practice, to the author's knowledge, though double walls separated by vacuum seem quite feasible for small devices such as ear-defenders, control cabins in noisy shops, etc. Other barriers are mass, the more the better, and various absorbent surfaces for high frequencies. The trouble with air-borne noise is its power of flowing round barriers and through holes. Absorbent baffles and splitters are beneficial, especially if the path can be turned through two right-angles.

Readers will have noticed that on jet engines, where baffles would interfere with the thrust, exhaust noise is attenuated by subdividing the jet, causing it to contact the free air over as large a periphery as possible. At the inlet, noise is reduced by minimising unevenness in the flow, for example, by omitting inlet guide vanes before the first compressor. Propeller noise is kept within bounds by mini-mising shock-waves originating from the propeller tips, partly by restricting the speed and partly by the profile, and also by enlarging the blade surface to reduce the unit pressure, though at some sacrifice of efficiency.

Two surfaces with an air gap between them attenuate noise quite well, for example, double-glazed windows and 'false' or 'suspended' ceilings carried on beams not connected to the floor above. Any rigid connection between the surfaces would produce a sounding-board effect. Since the cavity and the separate panels each have their own resonant frequency, the space can with advantage be filled with soft, spongy material.

Amini *et al.*[36] describe a sandwich material, available commercially, consisting of two metal sheets joined by a layer of soft plastic to provide flexural (shear) damping while maintaining good strength and stiffness in bending.

# 10

# Some Points on Manufacture and Appearance

Designing to ease manufacture must be based on a knowledge of the processes in existence and more particularly on those available in the firm or by contract, in other words those likely to be used in practice, though the purchasing of plant for a new product is not necessarily ruled out.

The cheapest process, generally speaking, is pressing and drawing, using strip material. Advice here starts with material economy; figure 103a shows an obviously wasteful layout. Figure 103b uses 16 per cent less material. A slight redesign of the shape (figure 103c) pushes the saving up to 19 per cent. Strip width can be chosen fairly freely, being cut from wider strip by adjustable machines. Narrow residues can be used up by layouts such as in figure 103d, 8 per cent more economical than in figure 103a. Example (e)[61] shows a redesigned insulating spacer using the whole strip width, saving 30 per cent on material and using a simpler punch.

Strip tends to have unequal ductility and hardness lengthways and crossways which may restrict the choice of layout. Pressed parabolic reflectors can have unequal spring-back and come out slightly elliptical. For product consistency they should all be orientated in the same direction relative to the original strip, perhaps by a deliberate orientation tag designed into the blank.

One design point in blanking is the burr produced by the shearing action on the edges and around any holes (figure 104). Unless the design is arranged so that the burrs are harmless, they may have to be removed, for example, by barrelling with abrasive chips or by hand, adding to the cost of manufacture. Note that the burrs in the new design at (e) face one up, one down. Another point about blanking is the minimum size of holes and minimum width of the component. These are limited by weakness of the punch which is generally of the same outline as the part, and also by difficulty in holding the material down adequately. Preferably hole diameter and least-part width should be greater than the thickness.

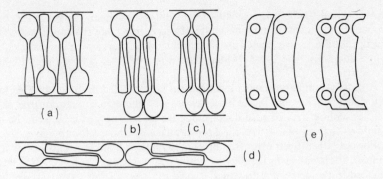

(a)
(b)
(c)
(d)
(e)
Figure 103. Blanking economy

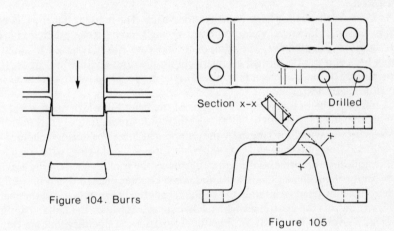

Figure 104. Burrs

Section x-x
Drilled
Figure 105

Figure 106

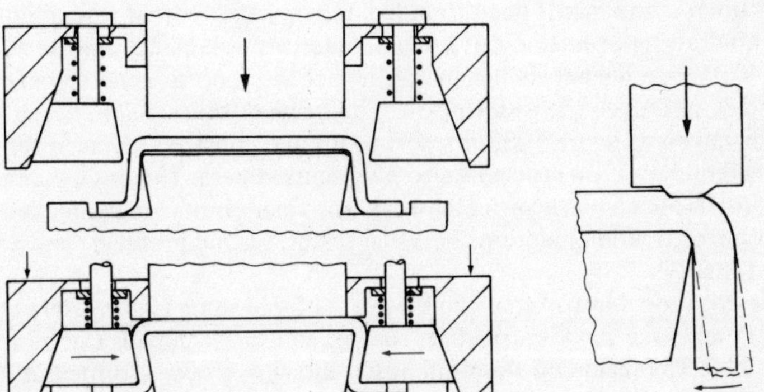

An example is the gearwheel on a light-duty boat winch. To make this by blanking, the thickness must be limited to something like the tooth width; this is not strong enough, so two blanks are used welded together, which is clearly cheaper than a single machined gear. If quantities warranted it, a sintered gear might be competitive.

A more accurate edge with less burr can be produced by precision-blanking, using very sharp punches and dies with no clearance. The punches cannot pass through the dies without risk of edge damage but are set to stop just level with the die edge. It follows that more massive, rigid presses have to be used for any given size of job.

Forming operations can often be combined with blanking. This makes for more complex tools but saves handling. In progressive work the component is left partly attached to the original strip which is used to carry it from stage to stage until the final parting-off. Forming divides broadly into bending and drawing, which do not greatly change the material thickness, and the more drastic operations of forging and extrusion. Rolling, wire-drawing and tube-making are outside the scope of this discussion.

In bending the main design point is spring-back. The amount depends on corner radius, material thickness, yield point and Young's modulus. Its prediction is dealt with in presswork manuals. The component in figure 105 is designed to allow for spring-back and can be formed by relatively simple press-tools. If 90° or acute corners were needed, overbending or local swaging could be used, with tools as shown in figure 106.

A minor problem is section distortion. At the bend the width increases at the inner corner, decreasing at the outer, together with a reduction of thickness. In consequence, the midline length of the component increases during bending.

Tobacco tins, hub caps and car bodies are formed by drawing, with considerable tensile and shear deformation. Figure 107 shows the tooling for drawing a cylindrical cup, comprising a punch, a die and a blankholder pressed down on to the blank to keep it from wrinkling due to the tangential compression. As the material is drawn towards the centre, a slight thickening accompanies the radial lengthening. The ears shown are the result of directional variations in ductility and are trimmed off afterwards. The amount of deformation depends on metal ductility but it is rare to achieve more than the amounts shown in figure 107 in one operation. Rectangular components need rounded corners, each corner resembling one-quarter of a cup-drawing operation but a slight benefit is obtained from the adjacent flat parts. To obtain deeper shells one can repeat the process, generally after annealing the metal, or reduce the wall thickness by ironing (squashing between punch and die), as shown in figure 108, or resort to jointed construction as in food cans. Metalbox technology is too specialised to be discussed here. The basic seams used may be found in books on sheet-metal work, the finer points are connected with the best ways of treating junctions between seams, sealing problems and adaptability for automation.

The edges of thin material can be treated by beading (figure 109) to add strength and take away sharp edges. Bulges and screw-threads can be formed as in figure 110. The plain cup stands in a split die with grooves formed in it. The punch carries a rubber slug and a metal foot (in some cases the punch could be all rubber). As the punch descends, the metal foot is stopped by the base and the rubber, which is virtually incompressible, flows towards the grooves in the die, pushing the metal

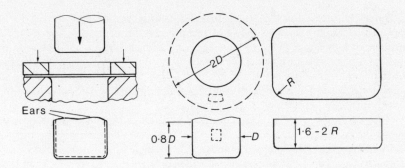

Figure 107.  Drawing cups and shells

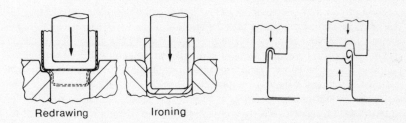

Redrawing        Ironing

Figure 108                    Figure 109

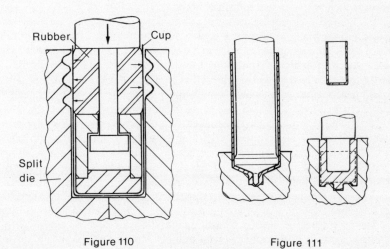

Figure 110                    Figure 111

before it. On retraction the rubber resumes its original diameter and the die halves open to release the finished article. A recently developed process uses oil instead of rubber for bulging, at the same time applying a down-force to the free edge of the cup[62]. A down-force may well also be useful in rubber bulging. The advantages of both oil and rubber are combined in the A.S.E.A. process[63] which uses a rubber-and-fabric capsule filled with oil to perform the displacing. The fabric is of mesh construction to permit large shape-changes.

General forging-processes are described in terms of the machine arrangement as press-, drop-, petro-, high-energy-rate, etc. For severe deformations the metal is worked hot. Any solid shape which can be extracted from a simple pair of dies seems to be feasible but undercuts and side-cores are not. Large tubes can be forged over a mandrel, small billets can be pierced, enlarged and forged or rolled into rings.

In dieforging, to secure thorough filling of the die shape excess material is provided and the surplus is squeezed out as flash, to be trimmed off. In extrusion, which is perhaps the most extreme case of forging, the flash is part of the design. Examples shown in figure 111 include a toothpaste tube, a zinc battery shell and a more general steel forging, with the original slug superimposed.

By using a roller, single-point deformation can be applied to flat material; if the thickness is unchanged this is called spinning; if the force applied is enough to thin the material as well as shape it, the process is called flow-turning.

One process for making parts of elaborate shape economically is to press powdered material to the required shape and pass this through a furnace where the particles are sintered (nearly melted) together. A slight allowance is made for shrinkage or, with some mixtures, growth. This process has been used for a long time to make tungsten rods for drawing into lamp filaments and for hard tool-tips, later extending into porous bearings, filters, etc. (A porous plug is sold as a silencer for exhaust air from pneumatic cylinders.) More recently high compaction pressures and accurate furnace-temperature control produce quite complex products with sufficient accuracy for immediate use, for example, timing chain sprockets complete with keyway. In other cases press-squeezing (sizing) or a light machining operation are used to finish both the most critical dimensions and also difficult features such as undercuts and small holes.

The compacting process does give rise to design limitations, but these are not severe and are likely to become even less so in future with vibratory or multi-stage processes. Compaction takes place in a die, usually open at the bottom as well as at the top, with punches coming in at both ends. Figure 112a shows a die for partly spherical bushes. Fully spherical shapes are difficult since the punches would have feather-edges which would suffer under the heavy forces needed to compact the powder. A die for a flanged bush is shown in figure 112b. The body and flange need several separate, concentric punches to give a uniform ratio of compression to each part. Any design that gives rise to thin, weak punches is un-desirable, as are designs with sharp corners in plan view which require sharp-cornered punches and dies. Apart from this, non-circular work is quite feasible. In strength the product can be comparable with a casting of the same material. Mixtures are feasible, for example, copper—carbon brushes, graphite and bronze for bearings. Porosity can be arranged to order, from about one per cent upwards.

The most versatile process is casting, in moulds of sand or ceramic with various binding agents or in permanent metal moulds. Moulding of plastics is almost as

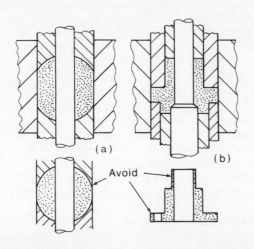

(a)

(b)

Avoid

Figure 112

versatile. Detailed advice on casting design must be combined with general policy. It is often useful to combine all the complications required into one or a few main castings, leaving the simpler parts be be made from solid bar or forgings. The degree of split-up is chosen for ease of casting, machining, assembly and replacement. In very large jobs, the split-up may be governed by melting capacity or crane-handling facilities.

Machine tools divide naturally into beds, heads and columns. Large pumps and turbines split on the centre-line, small ones break up into main bodies and volute casings, assembled axially. Hydraulic presses are made from cross-heads and tie-rods, electric motors have centre-bodies and end-frames. Farm and textile machines tend to resemble lawnmowers, with cast end-plates and light cross-bars. A split main body may simplify casting but detracts from rigidity and requires more machining than a one-piece body with access holes, as for example in figure 113a. Some large engines are made by welding together cast steel sub-units, thus combining ease of casting with a rigid result.

In large castings the main aspect is successful metal flow, cooling rates and grain structure. In small repetitive castings ease of moulding becomes important. Large castings are liable to be coarse-grained and so not amenable to fine surface finish. Hydraulic presses are often designed with rams and return rams (figure 113b). The cylinder surface need not have a good finish except at the seals and never wears out. Wear is concentrated at the bronze sleeves S and on the cheaper, smaller ram body which moreover is easily removed for refinishing.

Particularly difficult shapes can be cast by the ancient lost-wax process in which a full-size pattern is made and destroyed for each casting produced. For small parts this is of wax, surrounded by a ceramic plaster-cast, with wax feeder-ducts and air vents. The wax is melted out and molten metal poured into the cavity, giving excellent accuracy and surface finish. For larger castings the replica is made of low-density polystyrene foam. This too is surrounded, usually by sand, but is not melted out. Instead it is vaporised by the molten metal which takes its place. Toxic fumes are evolved but the mass of polystyrene involved is quite small. Foam could also be used in conjunction with the usual wooden reuseable patterns to form some of the undercuts, instead of loose pieces difficult to withdraw.

Pipes can be cast in a rotating mould, the molten metal being held circular by centrifugal force so that no core is needed. The process is not suitable for all alloys since the centrifuging acts as a powerful segregator, the densest phase migrating to the outside. This effect is made use of in lining large journal bearings; the harder cubical tin crystals collect at the inner surface to form the desirable high-spots, leaving the denser, cheaper lead-rich phase to form the soft matrix, thus cheapening *and* improving the final product.

Some detail aspects of casting are illustrated in figure 114.

(1) Avoid sharp concave corners (figure 114a) which encourage poor crystal structure just where stresses could be highest; moreover, sharp concave corners on the object mean sharp convex corners in the mould; in a sand mould, the sand will wash away here and finish up in the metal; a metal mould will erode and crack due to rapid heating and metal flow.

(2) Keep an even wall-thickness and avoid thick lumps (figure 114b) for since the melt must be hot enough to fill the thinnest section (c), it will cool too slowly in

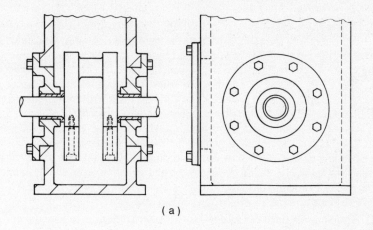

( a )

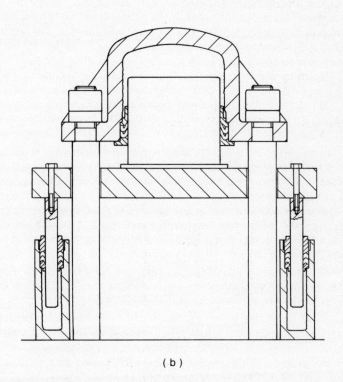

( b )

Figure 113

93

thicker parts, give coarse grain and possibly depressions or cavities (d, e). In high-pressure diecasting this limitation is much less severe than in sand or low-pressure diecasting.

(3) Help withdrawal of pattern or casting. In sand casting the pattern must withdraw from the mould without disturbing the sand; the casting is withdrawn by breaking up the sand mould. In metal dies there is no pattern problem but the casting has to be ejected successfully, any metal side-cores being withdrawn first. The casting tends to shrink on to interior cores. The designer must visualise where the mould will split so as to provide draft (taper) to aid withdrawal. Moulds generally consist of two main halves, preferably with a flat parting plane (figure 114f right) although stepped or curved parting surfaces can be tolerated. Sand cores need locating points (prints). Some undercuts are possible but need complex cores with loose or sliding pieces which slow down production. The student is advised to cut up a diecast engine piston and deduce the core formation from the flash-lines or witness marks. Some pistons are electron-beam welded to avoid undercuts and also to allow different metals to be used for crown and skirt.

(4) Trimming off flash can affect design. External flash is to be expected at any parting line; it is particularly severe on pressure diecastings. Flat parting planes and convex outlines help in flash removal. If the outline is circular a light lathe operation may be appropriate, giving a better finish than hand-trimming. The largest flash and the hardest to remove comes at junctions between cores. In sand casting the design may have to specify one-piece cores for valve bodies, pipe junctions, etc. In small diecastings the de-flashing may demand a drilling operation; if so it may be possible to do without some particularly deep or awkward cored holes, merely casting-in the starting dimples. If a high polish is required on the parting line, a slight ridge is helpful; smooth spherical surfaces make minor imperfections conspicuous. A scalloped edge is another way of disguising the flash-line (figure 114g).

(5) Sink marks (figure 114d) are slight depressions noticeable in flat surfaces on diecastings and plastic mouldings. They are due to delayed shrinkage at the base of bosses or internal ribs and can be disguised by external styling features such as pimples or ridges. Figure 114h shows an example where a double benefit is obtained. The component shown is nearly square and this shape tends to be visually unsatisfactory; the feature lines improve the apparent shape. Note also the slight convexity which minimises shrinkage stresses and the corner treatment for access to the screw-holes, saving over-all size and material.

(6) Lettering is usually upstanding. In sand-casting metal characters are screwed to the pattern; in diecasting they are engraved in the mould, and are often surrounded by a raised frame for protection during handling.

The basic machining processes are assumed to be known to the reader. When designing, over all and in detail, it is important to know the general scale of production envisaged and the equipment likely to be available, in-house or by contract. For the purpose of this section we shall distinguish four typical bands

(1) one-offs and tool-making with a cycle time of weeks or months,

(2) the five to fifty per week band,

(3) the consumer durables produced at approximately one per minute per plant, and

(4) the bits and accessories band at perhaps ten per minute.

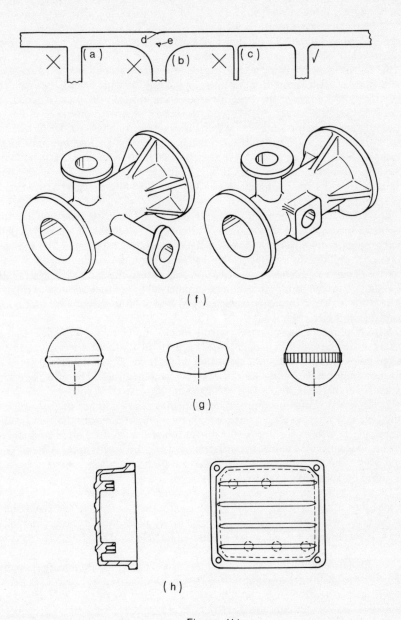

( f )

( g )

( h )

Figure 114

Band 1 obliges the designer to play very safe since the article generally has to be right first time. Band 2 may well allow a little prototype work so that the designer can be slightly more adventurous. In this band, numerically controlled machines could be particularly useful. Bands 3 and 4 require maximum co-operation between detail design and detailed production planning. Band 3 generally depends on existing processes but justifies purpose-built production lines, while band 4 begins to justify trying out or even inventing and developing processes new to the particular field.

In circular work, apart from avoiding obviously difficult operations, the detail designer should consider the available stock sizes; this can save much material and cost. If bright (scale-free) stock is used, the biggest diameter should be either of stock size or just below, enabling a light clean-up cut to be taken. If cheaper black bar, with mill-scale or a rough pickled surface, is used, the allowance must be sufficient to get below the rough layer and also for out-of-roundness. To calculate actual amounts, consult the works, even if only to increase goodwill. If the design demands a large head and a small long shank consider using an upset forging or a welded assembly.

If spanner-flats are required it is often best to use hexagon bar, though with expensive material such as brass it may be cheaper to use round bar and to design smaller spanner-flats to be made by hexagon-generating attachments or even by subsequent milling, or by substituting a tommy-bar hole.

With regard to circular work, attention is drawn to the somewhat less familiar processes of thread-milling and general circular-milling. Stock removal is rapid because work is shared by many cutting edges. It should be particularly useful on broad form-cuts and interrupted cuts.

Main bodies such as gearboxes, engine blocks and housings generally require milling, boring and drilling operations. It is particularly helpful to include in the design holes or lugs for holding down the body; these should be clear of any cuts. In production bands 3 and 4 any faces to be machined should be on a common outline, even if elaborate, so that one gang-milling operation suffices.

At this point it may be appropriate to mention the existence of copy-millers which make a copy, in steel or whatever is required, of a master built up from wood, plaster, etc.; moreoever some have a mirror-image facility for paired components. Although essentially tool-room machines they may be useful in band 1 and prototype work. The mirror-image facility is also available on numerically controlled machines by a sign-reversing switch on one main axis.

Large-quantity boring generally uses standard boring-heads; in band 2 they will be set up to finish the component in one or at most two set-ups and the designer can be most helpful by considering access, angles all in one plane, etc. In bands 3 and 4, transfer lines are likely to be employed; these tend to give more freedom of choice.

The reader is no doubt aware of the use of broaching for producing splined holes, straight or helical, keyways and other grooves. One feature of broaches is the use of several rows of teeth identical in size, so that the later teeth take over as the leading ones wear and correct sizes are produced for a long time. Sometimes broaches are used (economically) for external contours or flats. Designs needing special broaches would generally be worth considering in bands 3 and 4, more rarely in band 2.

Drilling may seem too simple to mention; only two points are made: in casting design it is often possible to provide flat entry and exit points rather than sloping or curved ones which may embarrass or even break the drills. Drilled holes are slightly too rough and inconsistent for many purposes, for example, bearing surfaces, press-fits or dowel locations; they are improved by reaming or by ball-burnishing, forcing a hard steel or tungsten carbide ball through a slightly undersized drilled hole. This process is also suitable for improving the inner surface of drawn tubes.

Two electric machining processes in current use are suitable for complicated shapes, hard or awkward materials and flimsy workpieces. They are very amenable to automation. One is electrical discharge or spark machining (E.D.M.). An electrode, often of carbon, is fed towards the workpiece while a pulsed d.c. is passed between them. A flow of kerosene is maintained between the faces and the feed-rate is controlled by the voltage required to pass the current. This is kept high, ensuring that no short-circuit is formed so that the current produces sparks of metal vapour which condenses in the kerosene and is washed away by it. The process is rather slow but reproduces the electrode shape or outline with great accuracy. It has long been used in die-sinking and is one of the few ways of machining tungsten carbide wire-drawing dies. A blanking punch can be used as electrode to cut its matching die by E.D.M.

The other process is similar but uses an electrolyte between the electrode and work, passing a high current at low voltage. It is called electro-chemical machining, E.C.M., and is preferred for production work, since the rate of metal removal is high; however, accuracy of reproduction is slightly less than that of E.D.M. In both processes some wear of the master eventually takes place. Both processes are burr-free and suitable for delicate work, including long deep holes of any shape.

On appearance and styling only a short comment is offered, not enough for a separate chapter. The outer form of a device should give some indication of its function according to accepted custom; knobs for pulling should look and feel different from knobs for rotation; controls in constant use should differ visually from those for infrequent adjustments. This means that styling must not be divorced from ergonomics.

The style in which the designer expresses his intentions varies from age to age: in Victorian days elaborately fluted lamp-standards, delicately curved, not very rigid machine frames, and oil reservoirs like ducal soup-tureens in polished brass were considered the height of elegance; some seventy years later they were ridiculed, only to be appreciated again now. There is an ever-present wish to decorate, as witnessed by gipsy caravans, canal-folk's goods and peasant art generally which coexists with a desire for simple shapes and plain textures or natural effects like wood-grain and polished stone.

If a designer can recognise the spirit of the age as it affects his particular field, he will generally choose to (or be made to) design conservatively for the mass-market, reserving the *avant-garde* flights of fancy for the small-quantity luxury market. Surface finish is place-and-time-dependent: in some places and periods satin-finish is welcome, in others it is rejected as inferior, possibly unhygienic.

In styling small products, functions are so varied that no general rules can be given apart from sticking to simplicity and trying out a prototype. The following examples show what can happen when visual design takes precedence over function.

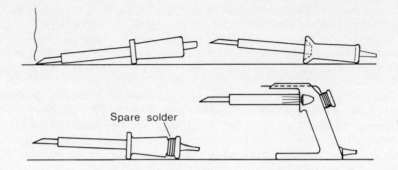

Spare solder

Figure 115

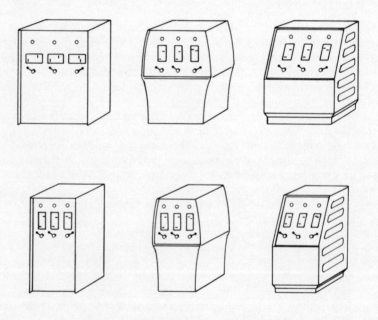

Figure 116

98

(1) A table knife with a heavy handle falls off the plate when the table is cleared.

(2) A telephone which is not heavy enough will be dragged across the desk by the flexible cord when the receiver is lifted, sliding loose papers out of place and possibly falling on to the floor if the cord has enough extension to allow this.

(3) A prize-winning soldering iron (figure 115) balanced so that the hot end is heavier than the handle results in burns on the work-bench. There are plenty of alternatives; a bigger handle, a change of shape, ballast or total re-design.

In all but the smallest objects the visual impression can be different from the true shape. A small but significant visual effect was known to the ancient Greeks. Doric columns look uniformly tapered, especially when close by, to achieve this they are slightly convex (barrelled). Similarly the upright sides of a traditional car radiator are slightly convex to make them look straight. The principle behind this rather odd effect could be argued as follows: when the eye meets a simple straight feature it starts looking for the ends, thereby underestimating the importance of the plain parts. The slight 'beefing-up' just compensates for this.

The designer can emphasise the vertical or the horizontal features by steps, slopes, ridges, fluting, trim or bands of colour. Stout people avoid wearing clothes with horizontal stripes but in machinery dominant horizontals give an impression of stability and confidence while verticals are dramatic and visually top-heavy. If verticals are unavoidable they should get narrower towards the top.

Figure 116 aims to show something of the interdependence of shape and styling. It is of course realised that

(1) This is a very subjective matter.

(2) Simple sketch presentation is only part of the story: the designer can deceive himself and others by presenting his favoured solution in a better light than the rest.

(3) A great deal depends on detail treatment such as corner radii or bevels, junctions between panels, fasteners, surface texture, colour contrasts, bright trim, etc.

The examples assume that over-all height, panel height and the approximate volume are common, being dictated by the equipment to be accommodated. Subject to the remarks above, it seems fair to say that the cheap plain solution is worth considering except in crowded spaces where the other variants provide useful toe-room. In practice the cheap version tends to look crude unless improved by sophisticated detail (which makes it less cheap). The upright pedestal treatment looks fine when narrow but matronly and slightly ridiculous when broader. The sloping-face model, often the most satisfactory functionally, looks solid and stable in the broad model but rather pinched in the narrow version.

# Appendix

## List of symbols

Symbols are defined locally in the text but this list is provided for convenience, and is arranged roughly in order of occurrence in the text.

| | | |
|---|---|---|
| $E$ | = | Young's modulus |
| $G$ | = | shear modulus |
| $\nu$ | = | Poisson's ratio |
| $w$ | = | load per unit length |
| $F$ | = | force (usually in a component) |
| $P$ | = | load (usually imposed on a structure) |
| $p$ | = | pressure (force per unit area) |
| $\rho$ | = | density |
| $\sigma$ | = | stress, usually tensile |
| $R$ | = | radius, and sometimes reaction |
| $D$ | = | diameter, and occasionally depth of a beam |
| $d$ | = | diameter, and occasionally displacement |
| $L$ | = | length |
| $t$ | = | thickness (wall thickness where appropriate) |
| $I$ | = | second moment of area |
| $M$ | = | mass, and sometimes bending moment |
| U.T.S. | = | ultimate tensile strength (= failing load/original C.S.A.) |
| C.S.A. | = | cross-sectional area |
| S.C.F. | = | stress-concentration factor |
| I.C. | = | Instantaneous centre |
| $\theta, \phi$ | = | angles |
| c.g. | = | centre of gravity (mass centre) |
| $\alpha$ | = | coefficient of linear thermal expansion, (expressed as p.p.m./°C) |
| $k$ | = | thermal conductivity in SI units, W/m °C, i.e. heat flow in watts per square metre C.S.A. under a temperature gradient of 1 °C per metre. |
| M.P. | = | melting point |
| $r$ | = | electrical resistivity, CGS version, $\mu\Omega$ cm, the resistance in microhms per cm length of a conductor of 1 cm$^2$ C.S.A. This old-fashioned CGS unit is used because electrical conductors of over 1 m$^2$ in cross-section are rare |

100

| $c$ | = | speed of sound in solid rods (bar velocity) |
| r.p.m. | = | revolutions per minute |
| Hz | = | hertz (cycles per second) |
| a.c. | = | alternating current |
| d.c. | = | direct current |
| $K$ | = | stiffness factor in torsion |
| $\delta$ | = | deflection |
| F.C. | = | flexural centre |

In chapter 3 expressions are dimensionless and can be used with any consistent units. In chapter 6 SI units are used: $MN/m^2$ for $E$, U.T.S. and stresses generally except where stated otherwise. The British Standard preferred unit of $N/mm^2$ gives the same numerical values as $MN/m^2$.

## A few conversion factors

1 standard atmosphere = 1013.25 mbar = 1.01325 bar = 101325 $N/m^2$
$\qquad\qquad\qquad\quad$ = 0.101325 $MN/m^2$ = 14.6959 $lbf/in.^2$
1 $tonf/in.^2$ = 2240 $lbf/in.^2$ = 15.443 $MN/m^2$
1 kilopound = 1000 lb mass or 1000 lbf (in United States)
1 kilopond (kp) = 1 kgf = 9.81 N (in Europe)
1 tonne (metric) = 0.984207 ton avoirdupois (long ton)
$\qquad\qquad\qquad$ = 1.10231 U.S. (short) ton

## Twist—bend buckling

The following is a simplified presentation of the twist—bend buckling situation. Figure 117a shows an I-section cantilever. It should be noted that ideal cantilevers tend to come in symmetrical pairs; a practical cantilever is longer than it seems because the end-fixing has some elasticity.

A down-load $P$ carried by the cantilever is attached at a distance $h$ from the centre We imagine giving the free end a deliberate small twist $\theta$, which is shown greatly exaggerated.

Two effects appear, a twisting moment $Ph\theta$ and a sideways load component relative to axis A—A. This latter component causes a sideways deflection $\delta$, where

$$\delta \approx \frac{P\theta L^3}{6EI_{sid}}$$

where $E$ is Young's modulus for the material of the cantilever, $I_{sid}$ is the second moment of the section in sideways bending, about axis A—A. The factor is 6 rather than 3 since the twist varies from $\theta$ down to zero at the fixing.

This gives a further twisting moment of $P\delta$ at the end, an average along the length of $\frac{5}{8}P\delta$. The total twisting moment $T$ is given by

$$T = \tfrac{5}{8}P\delta + Ph\theta \tag{1}$$

Such a moment will produce a twist angle of $TL/(KG)$, where $K$ is the torsional stiffness constant of the section (see below) and $G$ is the modulus of rigidity for the material.

101

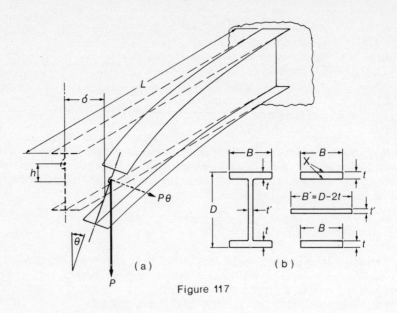

(a)

(b)

Figure 117

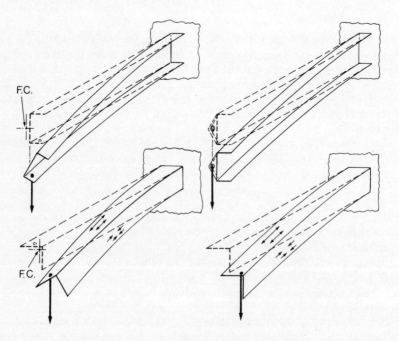

Figure 118

102

If we make the load $P$ so great that the angle of twist produced by it is equal to the imposed twist $\theta$, the twist becomes self-sustaining and mathematically indeterminate, indicating a buckling condition. In practice the twist may limit itself by the increase of $K$ which comes with large deflections producing substantial changes of length; on the other hand it may become catastrophic through localised kinking.

$$\theta = \left( \frac{\frac{5}{8}P \times P\theta L^3}{6EI_{sid}} + Ph\theta \right) \frac{L}{KG} \tag{2}$$

Hence

$$KG = \frac{5P^2 L^4}{48EI_{sid}} + PhL$$

$$= \frac{5P^2 L^4}{48EI_{sid}} \left( 1 + \frac{48hEI_{sid}}{5PL^3} \right) \tag{4}$$

Ignore the second term temporarily and find a temporary value of $P$, called $P'$, where

$$P' = \frac{3.1}{L^2} \sqrt{(KGEI_{sid})}$$

This is substituted in the second term of equation 4 to give

$$KG = \frac{0.1P^2 L^4}{EI_{sid}} \left[ 1 + \frac{3.1hEI_{sid}}{L\sqrt{(KGEI_{sid})}} \right] \tag{5}$$

From this we extract the value of $P$. If this is very different from $P'$, we recycle it. Otherwise we obtain

$$P \approx \frac{3.1}{L^2} \sqrt{\left[ \frac{KGEI_{sid}}{1 + (3.1h/L)\sqrt{(EI_{sid}/KG)}} \right]}$$

A much fuller solution gives a similar form with slightly higher numbers[64], so the present treatment errs on the safe side.

For a beam of length $L$ with a central load and its ends restrained against twist but not against side-bending the load may be as above but with a numerical factor of perhaps 16. It must be emphasised that these are not design loads but loads at which collapse is extremely likely. Standard I-beams could begin to fail in this way before orthodox failure once the span exceeds about $20D$, $30B$ or $100t$.

The torsional rigidity of I-beams, channels, etc., is substantially that of all the flat strips of which it is composed. For example, consider the I-beam shown in figure 117b. This consists of three bars and since for a rectangular section of width $B$ and thickness $t$

$$K = \tfrac{1}{3}Bt^3 \left( 1 - 0.63\frac{t}{B} \right)$$

the total is easily calculated.

Incidentally, the torsional shear stress in a rectangular bar is highest at point X, amounting to $(3B + 1.8t)/(B^2 t^2)$ times the torque applied to the bar which would be about one-third of the torque applied to the I-beam shown. At other points it is lower, in the inverse ratio of the distance from the centre. In the complete I-beam the bending stresses are also likely to be important.

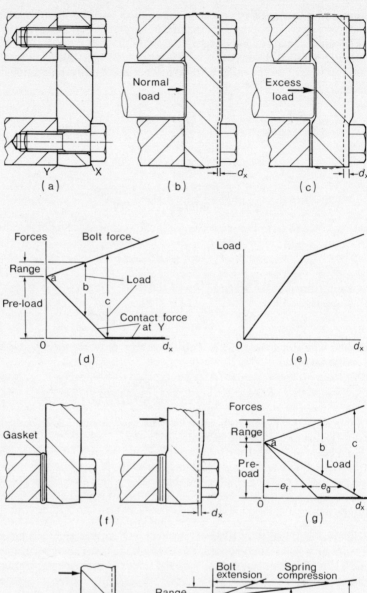

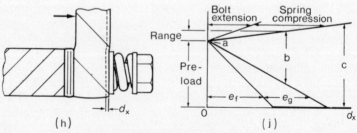

Figure 119

104

## Unsymmetrical sections

Unsymmetrical sections have two properties which can be helpful or otherwise. They tend to twist unless the load-line passes through the flexural centre or shear centre. This phenomenon is connected with the flow of shear round the section. Angle sections have the further property that, loaded in the usual way, even if through the flexural centre, they deflect sideways as well as in the load direction. The tension side gets longer and is offset relative to the compression side which gets shorter. The result is some lateral curvature (figure 118).

## Behaviour of bolted joints

Figure 119a shows a flanged joint, bolted and with initial tension. The effective stiffness of the bolts can be estimated; that of the flange and its surroundings is more difficult but will generally be substantially greater than the bolts holding it. Considering one bolt and its share of joint and loading: if a load is applied as shown be it mechanical or hydraulic, face X is displaced by an amount $d_X$, the bolt or stud elongates, and the flange relaxes and thickens, as shown (exaggerated) in figure 119b.

The load is resisted partly by the increased bolt force, partly by the reduced contact force at Y. This force cannot become negative, so eventually the faces part (figure 119c) and the bolt alone is effective. These relations are shown in figure 119d and the load—extension graph for the joint is plotted in figure 119e.

Figure 119f shows a joint with a gasket. Gaskets are usually required to prevent leakage of fluid and should be soft enough to settle into uneven gaps but also resilient enough to maintain a tight joint during changes of temperature, deflections due to load changes, creep etc. Figure 119g shows the effect of the same loads as in (b) and (c). The flange extension $e_f$ and gasket extension $e_g$ add up, making a larger total displacement of face X. The contact force at faces Y and Z has changed less than in (b) and (c), therefore the bolt-force must change more. For the same loads we now have a larger range of force and stress, which could matter if the load-cycle is repeated many times. The bigger displacement of X means that the structure is less rigid and may make more noise.

The bolt-force changes can be reduced by making the bolt more resilient, making it longer and slimmer, etc. (figure 79), or by use of a spring as in figure 119h. The assembly is even less rigid now so that face X deflects more than ever and separation occurs at a lower load than (c), but the bolt is treated more kindly since the displacement of X is taken up largely by the spring deflection. For the same normal bolt load, both the stress range and the stress at overloads are reduced (figure 119j).

If it is important to have a rigid assembly, the best arrangement would be a face-to-face joint with a resilient gasket set into a recess. This however is very demanding on the gasket material properties and is most likely to be successful in those cases where self-energising rubber seals can be used, supported by metal spring action.

## Metal bellows expansion joints

Flexible metallic bellows are used as pipe expansion joints, accommodating length or angle changes. The flexibility resides mainly in the flat walls. Several thicknesses (plies) of metal can be used to resist higher pressures without loss

105

of flexibility. The more corrugations, the more deflection can be allowed. For good fatigue life, makers state permissible extensions around 10 per cent and contractions up to, say, 15 per cent of the corrugated length. Angular movement is equivalent to extension at one side with contraction at the other; for convenience the permitted movement is expressed as an angle, perhaps $\pm \frac{1}{2}^{\circ}$ per corrugation. There is no obvious connection between the length and angle values but it should be noted that the proportions of corrugation pitch, inner diameter and radial width are relatively constant throughout any one range of designs.

Bellows are not only less stiff but also less strong axially then rigid pipes of the same diameter and pressure rating. In a pipe run as in figure 120a there is an obvious unbalanced force. If the pipes are not supported close to the joint, the bellows could overstretch and fail (figure 120b). In a straight run (figure 120c) there is no danger of this simple failure; however, the internal pressure acts on the flat walls of the corrugations, trying to extend the bellows and putting the bellows and the pipe between them into compression like a strut. This strut could fail as a fixed-fixed strut and pop out sideways, the energy coming from the supply pressure times the volume increase (Flixborough 1975). Stability of a single bellows has been analysed by Haringx,[65] that of a pipe run with two bellows by Newland.[66]

Sideways motion can be prevented without hindering the linear motion by strong guide-pillars or by an internal (or external) sleeve fixed at one end only. Axial limit stops can be included in many cases (see figure 120d). Bellows units can be purchased complete with restraining devices. This helps to prevent damage during transit or installation but the restraining devices supplied are not necessarily strong enough for severe service conditions such as pipe misalignment.

For large pipe movements the dog-leg layout can be employed (figure 120e). If there is enough room, pipe flexibility alone can be relied on to relieve the extension loads. Alternatively, bellows units can be used as hinges but must be protected from the unbalanced forces, preferably as shown in figure 120f – if the motion is sure to be in one plane only – or as in figure 120g, using a gimbal ring, if universal motion is needed. An internal sleeve may also be required if the corrugated length is large. Sometimes the swinging link is set up vertically where the pipeline crosses over a roadway.

### Some interesting and/or useful theorems

#### Maxwell's reciprocal theorem

In a structure, considerations of energy can be used to show that if a load at a point A produces a particular deflection at a point B, then transferring the same load to B will produce the self-same deflection at A. This is shown in many textbooks on structures; what is not shown is why we should be interested in the deflection at A due to a load at B. The deflections we most need to know are the maxima at any point, for clearance reasons, and the deflections at a load point. The latter are useful for resonance estimates and for assessing the influence of resilient foundations, etc. The maximum deflections at a point usually occur

106

when the load is at that point or not too far away; the main use of the theorem seems to be as an intermediate stage in calculating for rolling loads in redundant frames.

### Speed for maximum power from a belt drive

In a belt drive the maximum tension is limited by the fatigue strength of the belt in tension and bending. The ratio between tight-side and slack-side tension is limited by frictional grip considerations, as explained in most textbooks on the theory of machines, etc. The power transmitted would be proportional to the speed alone in any given set-up if it were not for centrifugal force in the belt. This theorem shows that the highest power transmitted in a given set-up occurs at that speed which makes the centrifugal tension one-third of the total permissible tension.

Calling the tight-side tension $T_1$ and the slack-side tension $T_2$ we suppose that the set-up is just tight enough to prevent slipping so as to minimise total tension (an exaggerated assumption), giving $T_1 = nT_2$ where $n$ is a ratio depending on the layout but not on the speed, $U$.

$$\text{Power transmitted} = (T_1 - T_2)\, U = T_1 U\left(1 - \frac{1}{n}\right)$$

The belt can only be allowed a certain maximum tension, $T_{max}$, which has to cover the driving tension $T_1$, a term for the bending which depends on pulley radii, $T_b$, and the centrifugal tension $T_c = wU^2/(g)$, $w$ being belt mass per unit length. $g$ is shown in brackets since it may not be needed, depending on the units system in which we are working. Thus

$$T_1 \leqslant T_{max} - T_b - T_c$$
$$\text{Power} \leqslant (T_{max} - T_b - wU^2/(g))\, U\left(1 - \frac{1}{n}\right)$$

To find the speed for maximum power, differentiate with respect to $U$ and set to zero

$$T_{max} - T_b - 3wU^2/(g) = 0$$

$$T_c = \frac{(T_{max} - T_b)}{3}$$

This speed could perhaps be realistic where very small pulleys for a given belt are used; in other situations the speed thus calculated is mugh higher than speeds usually recommended by belt manufacturers. Besides, the fatigue strength $T_{max}$ and values for $T_b$ are not readily available for proprietary belts — they must be derived backwards from catalogued power ratings. Finally it would be unwise to set up a belt drive so that the slack-side tension $T_2$ is only just sufficient.

### Constantinesco's theorem

This theorem closely resembles the preceding one mathematically; it shows the highest power that a given pipe can deliver from a source at fixed pressure. The pressure loss in a pipe tends to vary as the square of the flow rate. Thus if there is a supply at pressure $P$, the pressure at the outlet will be $P - kQ^2$ where $Q$ is the flow rate (please don't ask what to do with compressible fluids!).

$$\text{Power delivered} = (P - kQ^2)\, Q = PQ - kQ^3$$

$Q$ is the only variable in this system, so maximum power requires that

$$\frac{d\,\text{Power}}{dQ} = 0 = P - 3kQ^2$$

Thus for maximum power we must use that flow rate that makes the friction loss one-third of the supply pressure.

It is not to be supposed that a one-third loss is a generally sensible value for designing pipes to transmit power at steady rates. Nevertheless the theorem is worth looking at since there are some situations where peak power is needed infrequently and the pipe size is quite important. One such case is in aircraft, where hydraulic actuation of control surfaces and undercarriages is used and the weight of long pipe-runs needs minimising. Another possible application may be in small hydro-power schemes for farms or villages. These do not necessarily need expensive dams; the turbines may be improvised from boat propellers or second-hand centrifugal pumps running backwards. The water pipe from some convenient stream may well be the major expense.

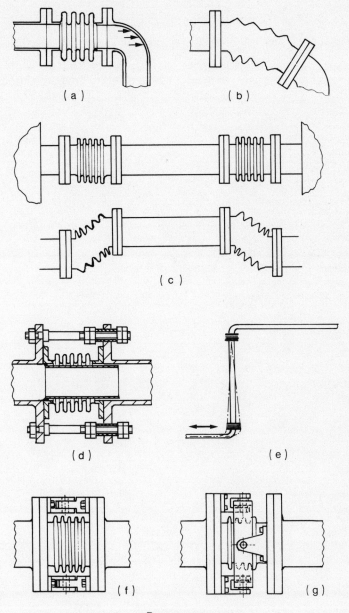

Figure 120

109

# References

1. Wesley E. Woodson and Donald W. Conover, *Human Engineering Guide for Equipment Designers* (University of California Press, 1964).
2. A. G. M. Michell, 'The Limit of Economy of Material in Frame Structures', *Phil. Mag.* 8 (1904) p. 589.
3. H. S. Y. Chan, *Optimum Michell Frameworks for Three Parallel Forces* (College of Aeronautics, Cranfield, Report Aero 167, 1960).
4. J. B. B. Owen, *Analysis and Design of Light Structures* (Arnold, London, 1965).
5. Production Engineering Research Association, *Survey of Literature on Machine Tool Structures,* pts 1 and 2 (Reports 166, 172, P.E.R.A., Melton Mowbray, Leics., 1967–8).
6. Engineering Sciences Data Unit, Data sheets 01.01.08, 01.01.09 (Structures) E.S.D.U., 251–259 Regent St., London W1R 7AD.
7. E.S.D.U. 04.06.01 (see reference 6).
8. E.S.D.U. 04.09.01 (see reference 6).
9. T. Von Karman and H. S. Tsien, 'The Buckling of Spherical Shells by External Pressure', *J. aeronaut. Sci.*, 7 (1939) p. 43.
10. American Society of Mechanical Engineers, *Boiler and Pressure Vessel Code,* section 8, (1971).
11. BS 1500 : Fusion welded pressure vessels for general purposes : Part 1 : 1958 Carbon and low alloy steels; Part 3 : 1965 Aluminium; 1500A : 1960 Carbon and low alloy steels.
12. A. H. Goodger, 'Fissuring along the Flow Structure of a Plate under Fillet Welds', *BSI News* (September 1966) p. 11.
13. P. Polak, 'Design Method for Corner Joints', *The Engineer,* 220 (1965) p. 155.
14. R. B. Heywood, *Designing against Fatigue* (Chapman & Hall, London, 1962).
15. P. G. Forrest, *Fatigue of Metals* (Pergamon, Oxford, 1962).
16. R. E. Peterson, *Stress Concentration Factors* (Wiley, Chichester, 1974).
17. R. Kuhnel, 'Axle Fractures in Railway Vehicles and their Causes', (in German), *Stahl Eisen,* 40 (1932) p. 965.
18. R. E. Peterson and A. M. Wahl, 'Fatigue of Shafts at Fitted Members', *Trans. Am. Soc. mech. Engrs,* 57 (1935) p. 1.
19. M. B. Coyle and S. J. Watson, 'Fatigue Strength of Turbine Shafts with Shrunk-on Discs', *Proc. Instn mech. Engrs,* 178 (1963) p. 147.
20. Motor Industry Research Association, *The Effects of Heat Cycling and Ageing on the Fatigue Strength of Fillet Rolled Components,* (Report 1962/5, M.I.R.A., Lindley, Nuneaton, Warks.).
21. A. M. Wahl, *Mechanical Springs* (McGraw-Hill, Maidenhead, 1963).
22. Civil Aircraft Accident Report, Comet G-ALYP 10.1.54 and Comet G-ALYY 8.4.54 (H.M.S.O., 1955).
23. T. E. Taylor, 'Effect of Test Pressure on the Fatigue Performance of Mild Steel Cylindrical Pressure Vessels Containing Nozzles', *Br. Weld. J.,* 14 (1967) p. 461.
24. R. M. Phelan, *Fundamentals of Mechanical Design* (McGraw-Hill, Maidenhead, 1962).
25. R. L. Wardlaw, *Some Approaches for Improving the Aerodynamic Stability of Bridge Road Decks* (National Research Council, Ottawa, 1972, DME/NAE 1972/2, 33).

26. J. O. Almen and A. Laszlo, 'The Uniform-section Disc Spring', *Trans. Am. Soc. mech. Engrs,* 58 (1936) p. 305.
27. E. F. Church, *Steam Turbines* (McGraw-Hill, Maidenhead, 1962).
28. J. S. Beggs, *Mechanism* (McGraw-Hill, Maidenhead, 1955).
29. S. B. Tuttle, *Mechanisms for Engineering Design* (Wiley, New York, 1967).
30. P. Polak, 'Prediction of IRS Roll-steer Geometry', *J. automot. Eng.,* 3 (1972) p. 48.
31. C. J. Smithells, *Metals Reference Book,* 3 vols (Butterworths, London, 1967).
32. G. W. Kaye and T. H. Laby, *Tables of Physical and Chemical Constants* (Longman, London, 1966).
33. J. Comrie, *Civil Engineers' Reference Book* (Butterworths, London, 1961).
34. J. L. Gray, 'Investigation into the Consequences of the Failure of a Turbine-generator at Hinkley Point "A" Power Station', *Proc. Instn mech. Engrs,* 186 (1972) p. 379
35. D. Kalderon, 'Steam Turbine Failure at Hinkley Point "A"', *Proc. Instn mech. Engrs,* 186 (1972) p. 341.
36. E. Amini and P. A. Atack, G. Pickard and R. F. Rimmer, D. K. C. Anderson, D. R. Cooper and J. Profit (A Series of Short Reviews on Duplex and Sandwich Metals and Composites) *Sh. Metal Inds,* 51 (1974) p. 1.
37. Manufacturers' literature, also J. Holden and W. Paton, R. Tetlow and G. H. Tilbury, *Proceedings of the Engineering Design Conference, Brighton, 1970* available from I.P.C. Mercury House Group, Mercury House, Waterloo Rd., London SE1.
38. CP 152: Glazing and fixing of glass for buildings.
39. F. J. T. Maloney, *Glass in the Modern World* (Aldus, London, 1967) p. 155.
40. BS 153: Steel girder bridges.
41. H. Thielsch, *Defects and Failures in Pressure Vessels and Piping* (Van Nostrand Reinhold, London, 1965).
42. K. F. Glaser and G. E. Johnson, *S.A.E. Jl.* 82 (1974) p. 21.
43. E.S.D.U. 67020, 68045, 69001
44. A. J. Phillips, 'The Design History of a V-8 Engine', *Proc. Auto. Div. Instn mech. Engrs,* 9 (1961–2) p. 339.
45. Ministry of Housing and Local Government, *Report on Collapse of Flats at Ronan Point, Canning Town,* London (H.M.S.O., 1968).
46. A. N. Gent, 'Elastic Stability of Rubber Compression Joints', *J. mech. Engng Sci.,* 6 (1964) p. 318.
47. *"Machinery's" Handbook, U.S.A.,* 11th ed. (Machinery Publishing Co., New York, 1954) p. 515.
48. R. E. Hatton, *Introduction to Hydraulic Fluids* (Van Nostrand Reinhold, London, 1962).
49. R. H. Warring, *Fluids for Power Systems* (Trade and Technical Press, Morden, Surrey, 1970).
50. M. J. Neale, *Tribology Handbook* (Butterworths, London, 1973).
51. J. L. Brodie, 'The Development of the De Havilland Series of Engines for Light Aircraft', *Proc. Auto. Div. Instn mech. Engrs,* 2 (1950–1) p. 65.
52. Production Engineering Research Association, *Hydrostatic Bearing System Design,* (Reports 134, 141, P.E.R.A., Melton Mowbray, Leics.).
53. P. Polak, 'Graphite-loaded Silicone Rubber', *Rubb. Plast. Age,* 50 (1969) p. 196.
54. E. Mayer, *Mechanical Seals* (Iliffe, London, 1972).
55. H. Hontschik and I. Schmid, 'The Seat as Connecting Element Between Man and Motor Vehicle' (in German), *Auto.-tech. Z.,* 74 (1972) p. 133.
56. A. B. Davey and A. R. Payne, *Rubber in Engineering Practice* (Applied Science Publishers, Barking, 1965).
57. C. H. Kindl, 'New Features in Shock Absorbers with Inertia Control', *J. Soc. automot. Engrs,* 32 (1933) p. 172.
58. British Patent 1095657 (D. A. Avner, Girling Ltd).
59. P. B. Lindley, *Engineering Design in Natural Rubber* (Natural Rubber Producers' Research Association, Tewin Rd, Welwyn Garden City, Herts. or 19 Buckingham St., London WC2, 1974).
60. A. J. Coker, *Automobile Engineer's Reference Book* (Newnes, Feltham, 1959).
61. The Plessey Co. Ltd. 'Rear Bank Insulator', *Value Engineering* (1970) p. 104.
62. D. M. Woo, 'Tube Bulging Under Internal Pressure and Axial Force', *Trans. Am. Soc. mech.*

*Engrs,* 95 (1973) p. 219.
63. I. Stromblad, 'Fluid Forming of Sheet Steel in the Quintus Press', *Sh. Metal Inds,* 47 (1970) p. 41.
64. R. J. Roark, *Formulas for Stress and Strain* (McGraw-Hill, Maidenhead, 1965).
65. J. A. Haringx, 'Instability of Bellows Subjected to Internal Pressure', *Philips Res. Rep.,* 7 (1952) p. 189.
66. D. E. Newland, 'Buckling of Double Bellows Expansion Joints under Internal Pressure', *J. mech. Engng Sci.,* 6 (1964) p. 270.

# Index